Aromatic character and aromaticity

Cambridge Chemistry Textbook Series

Aromatic character and aromaticity

G. M. BADGER

Vice-Chancellor,
University of Adelaide

CAMBRIDGE
at the University Press, 1969

Published by the Syndics of the Cambridge University Press
Bentley House, 200 Euston Road, London N.W.1
American Branch: 32 East 57th Street, New York, N.Y.10022

Library of Congress Catalogue Card Number: 68–29650
Standard Book Numbers:
521 07339 1 clothbound
521 09543 3 paperback

Printed in Great Britain
at the University Printing House, Cambridge
(Brooke Crutchley, University Printer)

Contents

Preface

Aromatic compounds have provided many intriguing problems for chemists. At first, attention was directed to benzene and its derivatives; and the term 'aromatic' was once regarded as synonymous with 'benzenoid'. It was soon recognized, however, that the structures and reactions of many unsaturated heterocyclic compounds are similar to those of substituted benzenes, and the former were therefore called heterocyclic aromatic compounds. They are now commonly called hetero-aromatic compounds. In the last two or three decades there has been in increasing interest in non-benzenoid unsaturated cyclic compounds having 2, 4, 6, 8, 10, 12, 14, 16, 18, 20, 22, 24, 26, 28 or 30 π-electrons. Many new non-benzenoid compounds, both monocyclic and polycyclic, have been synthesized and studied; and it must be asked whether these substances are aromatic? Physical methods for the study of cyclic compounds have been improved, and new techniques have been devised, so that new evidence is now available.

This book attempts to summarize the evidence relating to the nature, and to the electronic structures, of aromatic compounds. Like the other books in this series, it is directed to students in the final year of an undergraduate honours chemistry course, and to those who are beginning graduate work. The physical evidence relating to the shape and size of the benzene molecule is first summarized, and this is followed by the wave-mechanical picture of its electronic structure. The polycyclic compounds and the hetero-aromatic compounds are similarly pictured, and aromaticity is then defined in terms of the electronic structure.

The second chapter is concerned with the physical methods which can be used to determine whether an unsaturated cyclic compound possesses aromaticity. The remaining chapters summarize recent studies on non-benzenoid cyclic compounds, and discuss the extent to which these compounds possess aromatic character and aromaticity.

It is a pleasure to express my thanks to Dr Jillian Teubner, whose help in the preparation of this book has been invaluable. I am also greatly indebted to my friends and colleagues, Dr J. A. Elix, Dr G. E. Lewis and Dr T. M. Spotswood, for many helpful discussions. Finally, I am grateful to Dr K. Schofield, and to the officers of Cambridge University Press, for the skilled attention which has greatly simplified my task in the publication of the book.

G.M.B.

Adelaide, 1968

1. Introduction

1.1. Aromatic character. Benzene was discovered in 1825 by Faraday; he obtained it from the condensate of a compressed illuminating gas which had been prepared by decomposing whale oil at red heat. In 1834 Mitscherlich found that the same hydrocarbon can be prepared from benzoic acid by dry distillation with lime, and he showed that it has the formula C_6H_6. A. W. Hofmann and C. B. Mansfield discovered benzene in coal tar in 1845, but it was not until 1848 that Mansfield isolated relatively pure benzene from this source. Mansfield carried out the distillations in a glass retort with the thermometer in the liquid, and he wrote:

'It has been, perhaps, the tedium of the methods necessary to effect a separation of mixed hydrocarbons from each other, which has deterred experienced chemists from devoting their time to disentangling the oils here treated of: and, perhaps, to have conducted the innumerable distillations necessary for this purpose in a laboratory imperfectly furnished with gas and other conveniences, would have been a task too laborious to have been persisted in.'

Faraday named his compound 'bicarburet of hydrogen'. Mitscherlich named it 'benzin'; but Liebig criticized this name as implying a relationship to strychnine and quinine and recommended 'benzol' (German, öl, oil). In 1837 A. Laurent suggested the name 'pheno' from the Greek 'I bear light' in recognition of the discovery of the hydrocarbon by Faraday in an illuminating gas. The name 'benzol' became established in Germany, but in England and France the name was eventually changed to 'benzene' to avoid confusion with the systematic names of alcohols. Laurent's suggestion was never adopted for the hydrocarbon itself, but 'phenyl' has long been used to designate the C_6H_5 group.

Many benzene derivatives had been discovered long before the isolation of benzene itself. Benzoic acid was obtained from gum benzoin as early as the sixteenth century; and benzaldehyde (oil of bitter almonds), cymene (oil of carraway), and toluene (balsam of Tolu) have also been known for a very long time. The coal gas industry was developed in England during the first half of the nineteenth century. Coal tar was a troublesome by-product, but the successful isolation of benzene and of some benzene derivatives soon suggested possible uses for the material. Coal tar soon became the most important source of benzene and its derivatives.

Kekulé called benzene and its derivatives 'aromatic compounds' because of their characteristic odour. A special name was justified because it had become evident that benzene and its derivatives possess rather special properties, and contain a higher proportion of carbon than the fatty compounds. Benzene has the formula C_6H_6, but the saturated aliphatic compound with the same number of carbon atoms has the formula C_6H_{14}. On the other hand, benzene and its derivatives are not highly unsaturated substances, but are characterized by their remarkable stability. Benzene itself can be prepared by the high-temperature decarboxylation of benzoic acid, by distilling phenol with zinc dust, and even by passing acetylene through a red-hot tube.

In his classical papers on the constitution of aromatic compounds, August Kekulé (1865, 1866) proposed the well-known cyclic structure for benzene, and he suggested that the peculiar properties of the aromatic compounds are dependent on the properties of this ring system. Kekulé therefore equated the terms 'aromatic' and 'benzenoid', and this view persisted for many years.

The peculiar properties of benzene and of its derivatives led to the use of the expression 'aromatic character'. Aromatic character cannot be rigidly defined, but can be illustrated by the following observations:

(i) A characteristic property of benzene is its thermal stability, and its ease of formation by pyrolytic methods.

(ii) Reagents such as nitric acid, sulphuric acid, and bromine, attack benzene under suitable conditions to give substitution products rather than addition products.

(iii) Benzene is remarkably resistant to oxidation: it is not oxidized by cold alkaline permanganate or by nitric acid, for example.

(iv) The properties of substituted benzenes are often different from those of analogous aliphatic compounds. Aniline is less basic than aliphatic amines, and it reacts with nitrous acid in the presence of hydrochloric acid to form benzene diazonium chloride. Phenol is a stronger acid than the alcohols. Benzoic acid is a stronger acid than acetic acid. Halo-benzenes are much less reactive than aliphatic halogen compounds.

1.2. The structural formula of benzene. The problem of providing an adequate structural formula for benzene engaged the minds of chemists for many years. In 1865 Kekulé published his paper entitled *Sur la constitution des substances aromatiques* in which he suggested that the six atoms of carbon form a closed chain. In another paper published in the same year he used a regular hexagon formula in which the six carbon atoms were labelled by the letters *a–f*. The familiar hexagon formula with alternate single and double bonds was first used in a paper published in 1869 (Kekulé, 1869).

Kekulé showed that all disubstitution products must exist in three isomeric modifications, that there are only three isomers of the general formula $C_6H_3X_3$, and that there are six of the formula $C_6H_3X_2Y$. He was also able to explain the nature of the homologues of benzene and the essential difference between substitution in the nucleus and substitution in a side chain.

Although very successful in explaining the number and nature of the various substitution derivatives, Kekulé's cyclohexatriene formula (I) has always been regarded as only partly satisfactory: it is very difficult to explain the stability of benzene if it is assumed to have three ethylenic double bonds. From this point of view the structure (II) generally attributed to Dewar is even less satisfactory,

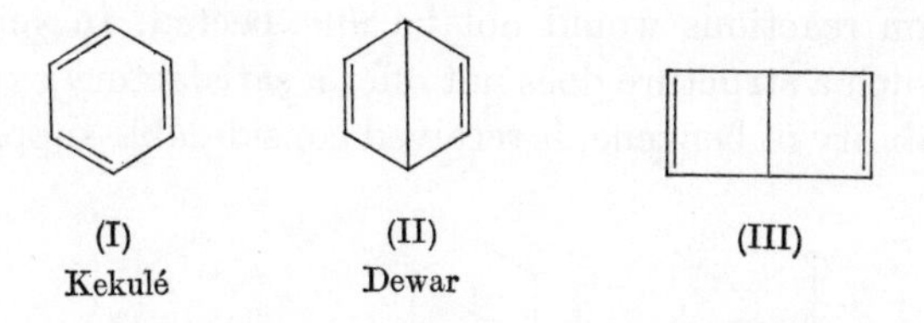

for it requires a long *para* bond. It is of interest that a compound (III) akin to 'Dewar benzene' has been synthesized (van Tamelen & Pappas, 1963), and substituted derivatives have also been reported.

One of the most serious objections to the Kekulé structure (I) was pointed out by Ladenburg. If the structure of benzene really involves alternate double and single bonds, then *ortho*-disubstituted derivatives should exist in two isomeric modifications (IV) and (V). Moreover, when the two groups are different, there

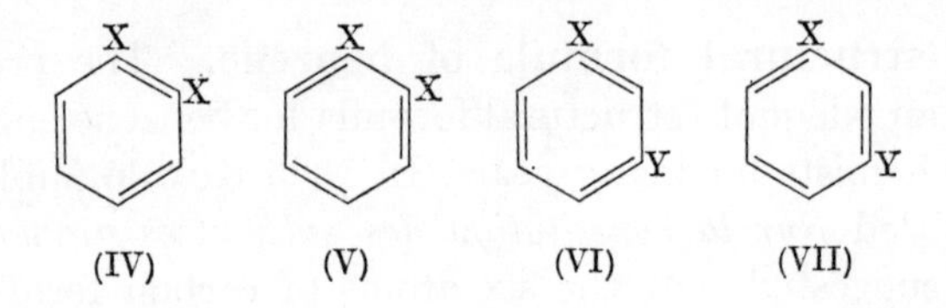

(IV) (V) (VI) (VII)

should also be two isomeric *meta* derivatives (VI) and (VII). As such isomers were, and are, unknown, some modification of the Kekulé structure appeared to be necessary. Victor Meyer suggested that the differences between such isomers might be so slight as to escape detection; but Kekulé further proposed that benzene has a kind of dynamic structure in which each carbon–carbon bond oscillates between a single and double bond. In other attempts to solve the benzene problem, Ladenburg tried to interpret its peculiar stability by picturing the molecule as a triangular prism (VIII); but this suggestion is only of historical interest for as well as being incapable of correctly representing relationships amongst benzene derivatives the prism structure is in conflict with the established planarity of benzene.

Claus also abandoned the use of double bonds and proposed a formula (IX) having three *para* bonds. In this structure, each carbon atom is supposed to be linked to three others, two in the *ortho* and one in the *para* position, so that *ortho–para* direction in substitution reactions would not be unexpected. In spite of the fact that such a structure does not offer a satisfactory explanation for the stability of benzene, it received considerable support at the time.

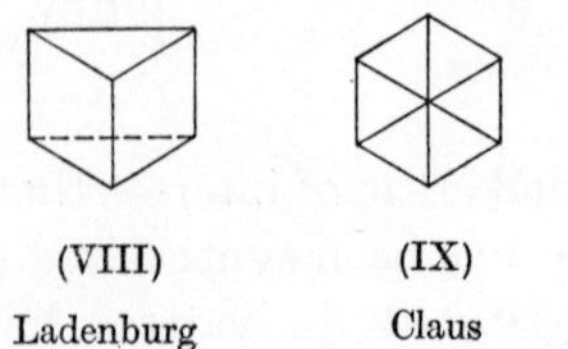

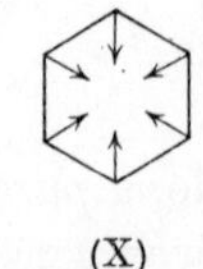

(VIII) Ladenburg | (IX) Claus | (X) Armstrong–Baeyer

Armstrong attempted to overcome the difficulty of the six unused valencies in benzene in another way (x). He suggested that 'the remaining six (valencies) react upon each other,—acting towards a centre as it were, so that the "affinity" may be said to be uniformly and symmetrically distributed'. Baeyer put forward a similar interpretation. These structures can hardly be seriously considered, for the hypothetical bonds directed towards the centre of the ring have no real meaning in terms of the modern electronic theory. The Armstrong–Baeyer formulation was little more than a pictorial representation of the fact that the problem of the fourth valency remained unsolved.

Of all the structures proposed for benzene that of Thiele (1899) is probably the nearest approach to the modern view. Thiele supposed that the reactivity of an olefinic linkage is due to incomplete saturation of the affinities of doubly-bound carbon atoms, which may be assumed to have partial valencies, represented by dotted lines. In conjugated compounds the partial valencies of adjacent carbon atoms linked by a single bond were supposed to neutralize each other with a resulting accumulation of partial valency at the ends of the conjugated system. According to Thiele, benzene is a conjugated system of double bonds *par excellence*, all the partial valencies being neutralized, as in (xi).

$>C=C<$ $CH_2=CH-CH=CH_2$

Isolated double bond Conjugated system

As the theory of conjugated systems developed, this formula was modified, and the fully symmetrical structure (xii) of J. J. Thomson, involving real fractions of covalent bonds, came into use.

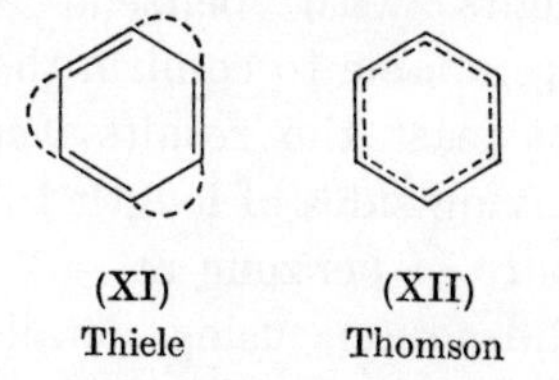

(XI) Thiele (XII) Thomson

1.3. The application of physical methods to the benzene problem.

X-ray and electron diffraction experiments. The structure of benzene has been established beyond doubt by the application of two types of physical method: X-ray and electron diffraction, and infrared spectroscopy.

In X-ray diffraction and in electron diffraction experiments, a picture of the molecule under investigation is built up by recombining the waves which it scatters. The electron diffraction method is most easily applied to gases and to volatile substances, and the X-ray method to crystalline compounds. In particular, the diffraction method lends itself to the accurate determination of interatomic distances, or bond lengths. By these means it has been possible to determine the normal covalent radii for a large number of atoms (Pauling, 1960). It was found, however, that the C—C bond length is not a constant, but varies considerably with the nature of the linkage. In diamond, and in the saturated aliphatic compounds, the C—C bond length has been found to be near 1·54 Å; in ethylene, the C═C double-bond length is about 1·34 Å; and in acetylene, the C≡C triple-bond length is 1·20 Å. In non-conjugated compounds these lengths are found to be virtually constant; but in conjugated compounds considerable variations have been observed, the formal double and triple bonds in such compounds being longer than normal and the formal single bonds being shorter than normal.

The C—C bond length in benzene, and in benzene derivatives, has been determined both by electron diffraction and X-ray diffraction experiments to be near 1·39 Å. For example, analysis of the X-ray diffraction pattern of crystals of hexamethylbenzene gave a value of 1·39 Å for the length of the aromatic C—C bonds; and the length of the bonds linking each methyl group to the annular carbon atoms was found to be 1·53 Å. Following electron diffraction experiments with benzene vapour, Pauling & Brockway (1934) were unable to confirm that all the C—C bonds are identical, but at least the results were consistent with a picture of benzene having sides of length 1·39 Å.

The crystal structure of benzene at $-3°$ has also been determined by X-ray diffraction, using detailed three-dimensional analysis (fig. 1.1). The benzene molecule was found to be a regular planar hexagon with C—C bond lengths of 1·392 Å. This value is in good agreement with the bond length (1·397 ± 0·001 Å) determined by electron diffraction of benzene vapour (Bastiansen, 1957).

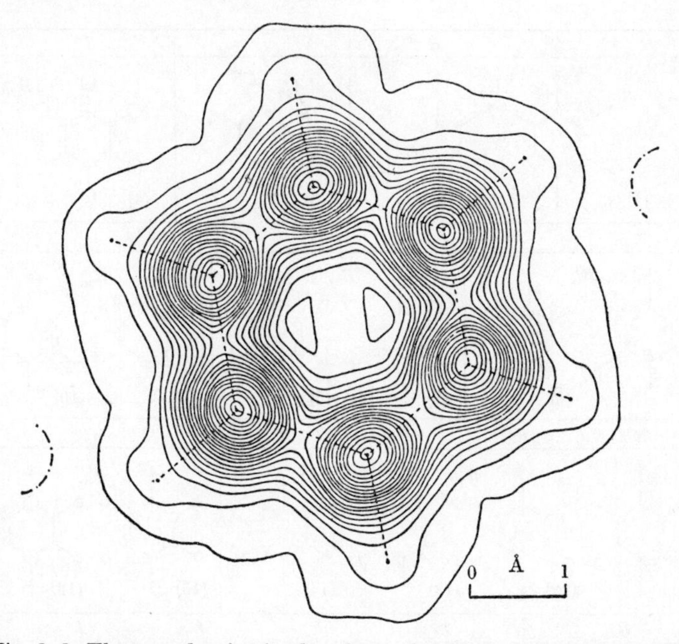

Fig. 1.1. Electron density in the plane of the benzene ring; contour interval 0·25e/Å^3 (from Cox, Cruickshank & Smith, 1958).

The physical evidence of electron diffraction and of X-ray diffraction therefore indicates that benzene has a regular planar hexagonal structure of side 1·39 Å.

Infrared spectroscopy. The application of spectroscopy to the study of the structure of benzene depends on the fact that a change from a lower to a higher energy state is accompanied by absorption of light, and a change from a higher to a lower energy state by emission of light. The frequency of the light emitted or absorbed is determined by the magnitude of the energy change involved, and is given by the equation

$$\Delta E = E_2 - E_1 = h\nu = \frac{hc}{\lambda},$$

where ΔE is the difference in energy between the two energy levels (E_2 and E_1), ν is the frequency of light absorbed or emitted, h is Planck's constant, and c is the velocity of light. Changes in rotational energy are associated with frequencies in the far infra-

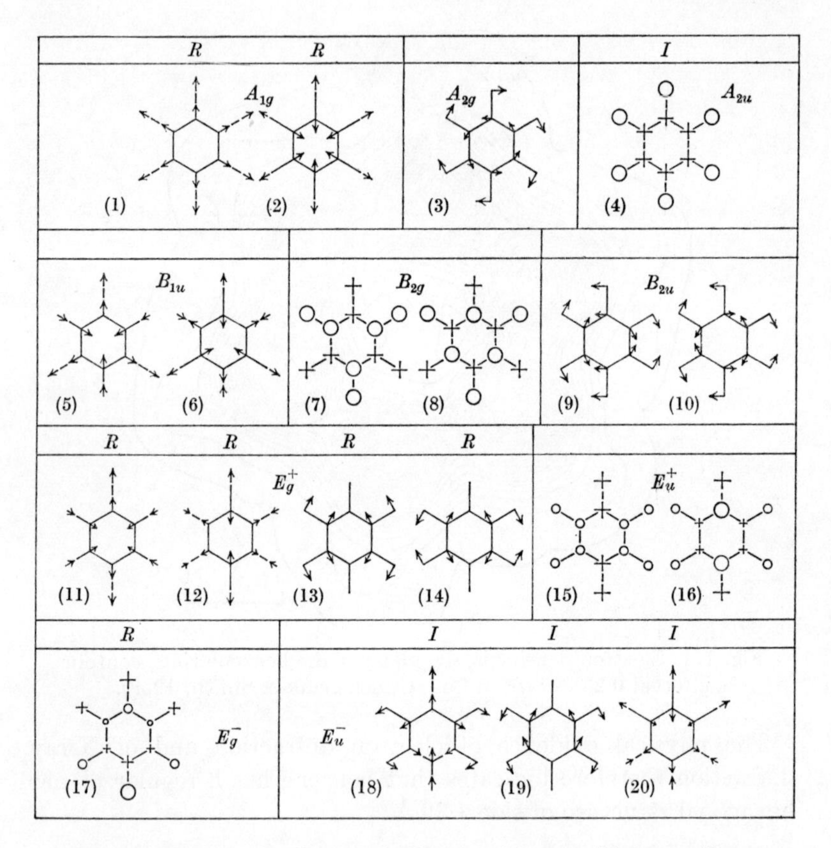

Fig. 1.2. Twenty fundamental vibrational forms of benzene (from Ingold, 1938).

red region; electronic transitions are associated with light in the ultraviolet and visible regions; and the vibration frequencies of molecules are in the infrared region.

Ingold (1938) has pointed out that if benzene has a planar, regular hexagonal structure it must have twenty fundamental vibrations, as illustrated in the accompanying diagrams (fig. 1.2). The arrows indicate motions in the plane of the ring, and the noughts and crosses motions perpendicular to the plane. Through a study of the spectral characteristics of benzene and of its deuterated analogues it was found possible to assign all the vibrational forms of the model to particular frequency bands in the spectra.

Only those vibrations which are associated with an oscillation of dipole moment are recorded in the infrared absorption spectra. The planar, regular hexagonal model for benzene has four such vibration forms (labelled I), and four fundamental absorption bands were observed in the infrared absorption spectrum of benzene vapour. Similarly, only those normal vibrations can appear as fundamentals in the Raman spectra which involve an oscillation of molecular polarizability. The model requires seven such vibrational forms, and seven fundamental bands were observed.

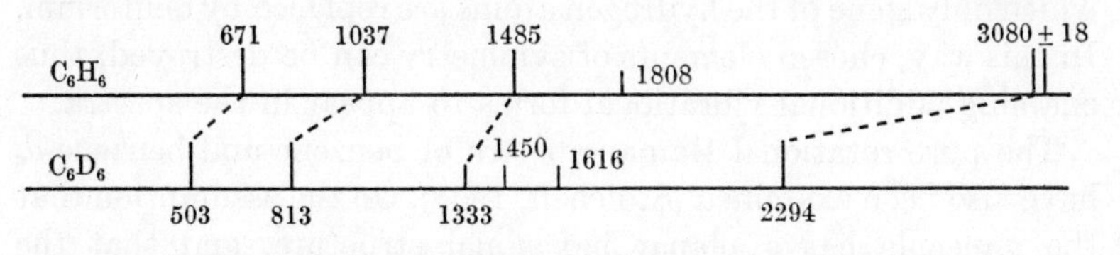

Fig. 1.3. Correlation of fundamentals in the infrared spectra of C_6H_6 and C_6D_6 (from Ingold, 1938).

The assignment of each band to a particular vibrational form was achieved by comparison of the spectrum of benzene with that of hexadeuterobenzene, C_6D_6; for the sole effect of the substitution of deuterium for hydrogen is to alter the *mass* and hence the vibration frequency (fig. 1.3). As the ratio of the masses was known, the frequency shifts for a given model were calculated, and the comparison of these calculated frequency shifts with those observed served the double purpose of assisting in or confirming the identification of the normal vibrations and of testing the model. Clearly the model which always gives the correct relations is proved.

The low-frequency band, 671 cm^{-1} in the infrared absorption spectrum of C_6H_6, and the 503 cm^{-1} band in C_6D_6 have been identified with the vibrational form in which the six hydrogen atoms together move perpendicularly to the plane of the ring (no. 4, fig. 1.2). Similarly, the lines at 849 cm^{-1} in the Raman spectrum of C_6H_6 and at 661 cm^{-1} in that of C_6D_6 have been assigned to the vibrational form in which the plane of the hydrogen atoms rocks over the plane of the carbon ring about an axis common to both planes (no. 17, fig. 1.2). The highest frequency line at 3062 cm^{-1} in the Raman spectrum of C_6H_6, and at 2292 cm^{-1} in C_6D_6, is the breathing vibration of the hydrogen atoms (no. 2, fig. 1.2). The frequency shift is large because almost all the motion is in the hydrogen (or deuterium) atom, whose mass changes in the ratio 1:2.

In this way eleven of the twenty vibrational forms of the model were investigated. The other vibrational forms are not recorded in either the infrared or the Raman spectra, and it was necessary to attack the problem in other ways. One method involved a study of the fluorescence (emission) spectra, and a second method involved the investigation of the Raman and infrared spectra of benzenes in which only some of the hydrogen atoms are replaced by deuterium. In this way, chosen elements of symmetry can be destroyed, thus enabling additional vibrational forms to appear in the spectra.

The pure rotational Raman spectra of benzene and benzene-d_6 have also been examined (Stoicheff, 1954). On the assumption that the molecule has a planar hexagonal structure, and that the interatomic distances are the same in both molecules, the C—C bond length was found to be $1{\cdot}3973 \pm 0{\cdot}0010$ Å, and the C—H distance $1{\cdot}084 \pm 0{\cdot}006$ Å. Finally, an investigation of the rotational Raman spectra of symmetrical benzene-d_3 gave a C—C bond length of 1·397 Å and a C—H bond length of 1·084 Å (Langseth & Stoicheff, 1956).

The successful conclusion of all this work has established the structure of benzene beyond all doubt. It is a plane regular hexagon of side 1·39 Å, and all the C—C bonds are entirely equivalent.

1.4. The application of wave mechanics. The electronic configuration of carbon in the ground state is $(1s)^2\,(2s)^2\,(2p_x)^1\,(2p_y)^1$. This configuration, with two unpaired electrons, corresponds to bivalency and it is, of course, well known that carbon rarely forms compounds in this way. The characteristic quadrivalent state can only be achieved if there are four unpaired electrons and Pauling (1931) described how this can be achieved by a process of exciting one of the $2s$ electrons into the empty $2p_z$ orbital, followed by hybridization.

When the four orbitals, $2s$, $2p_x$, $2p_y$, $2p_z$ are hybridized, the overlapping in phase occurs in such a way that four new orbitals are formed, all at angles of 109° 28′ to one another. This is termed tetrahedral hydridization, and the contours of ψ for such sp^3 orbitals can be drawn (fig. 1.4).

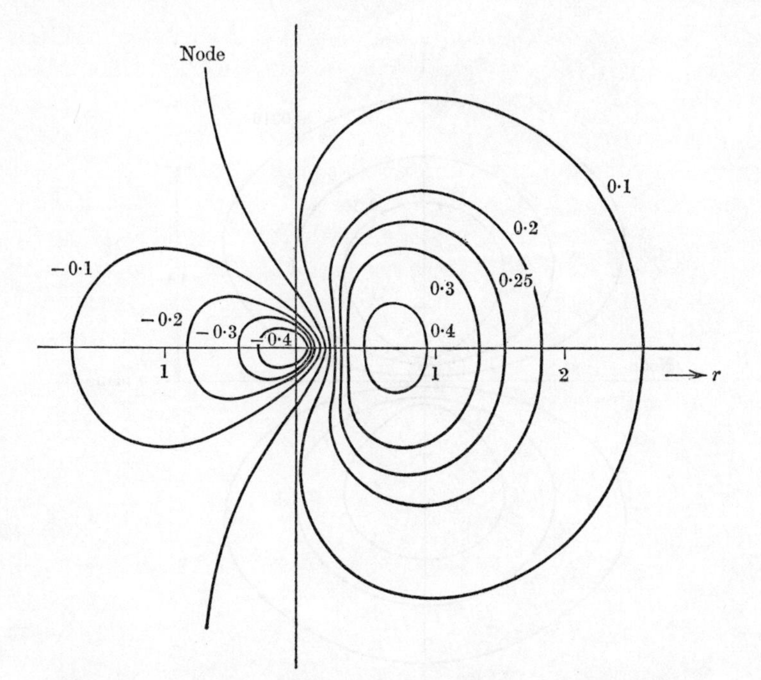

Fig. 1.4. Contours of ψ for a sp^3 orbital (after Coulson, 1952).

In trigonal hybridization, the $2s$, $2p_x$ and $2p_y$ orbitals are hybridized to give three equivalent orbitals in the xy plane. These orbitals are directed at angles of 120° to one another. The remaining $2p_z$ orbital (fig. 1.5) retains its directional character along the z-axis and is perpendicular to the plane of the other three orbitals. In digonal hybridization only the $2s$ and $2p_x$ orbitals are hybridized. Two new orbitals, which are orientated in opposite directions along a straight line, are obtained. The remaining $2p_y$ and $2p_z$ orbitals retain their directional character along the y and z axes respectively and are perpendicular to the orientation of the two hybrid orbitals and to one another.

When these orbitals are used for the carbon atom the nature of the bonds in carbon compounds becomes clear. In methane, each tetrahedrally hybridized orbital interacts with the $1s$ orbital of a hydrogen atom in such a way that a molecular orbital is formed. The principle of maximum overlapping indicates that the bonds are directed to the corners of a regular tetrahedron (Pauling, 1960), and this has been established experimentally.

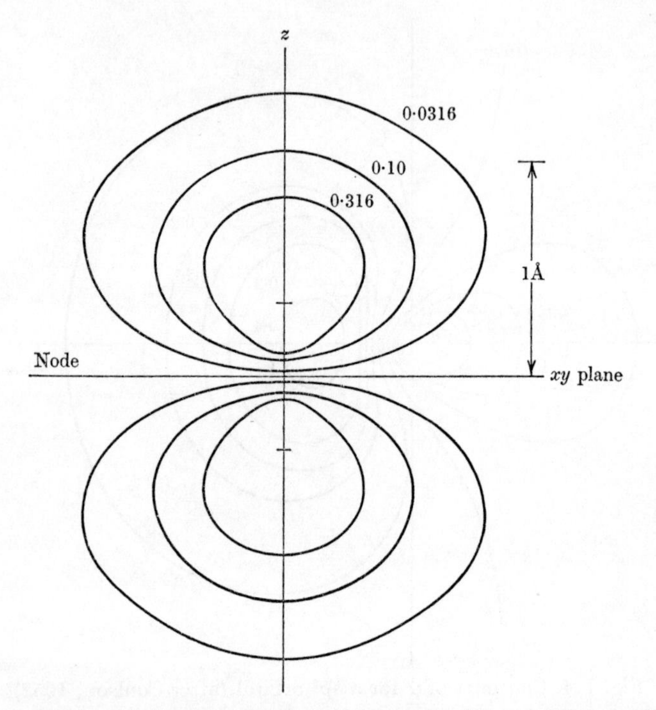

Fig. 1.5. Electron density contours of constant ψ^2 at 0·0316, 0·100 and 0·316 of maximum for a carbon $2p_z$ orbital (after Ogryzlo & Porter, 1963).

In ethylene, each carbon atom contributes three orbitals disposed at 120° to one another. Two of these are used for bonding to hydrogen atoms, and the third for bonding to the other carbon atom. The resulting C—H and C—C bonds are σ-bonds. Each carbon atom has a remaining $2p_z$ orbital, disposed in two lobes above and below the *xy* plane of the σ-bonds. These $2p_z$ orbitals overlap laterally to form π-orbitals. The resulting π-bond corresponds to one component of the classical 'double bond' in organic chemistry. The ψ^2 contour lines (fig. 1.6) show that the π-orbital has a node in the plane of the molecule; the boundary surfaces suggest two 'streamers', two 'buns' or two 'bananas' above and below the plane of the molecule. The same picture results if it is supposed that two sp^3 orbitals from each carbon atom interact to form the molecular orbitals (Pauling, 1959). This system clearly has considerable rigidity, and rotation about the C═C axis is restricted, so that *cis–trans* isomerism of suitable derivatives is possible.

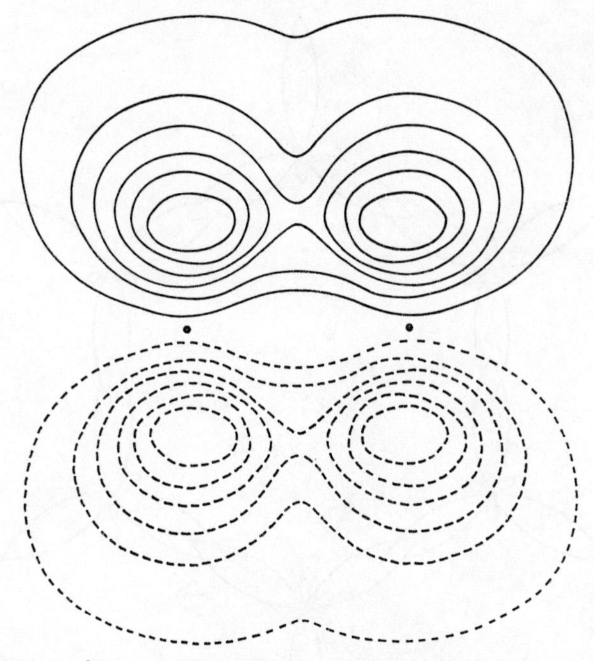

Fig. 1.6. ψ^2 contour lines for the π-electrons in ethylene (after Hückel, 1937).

In acetylene, each carbon atom contributes two digonal hybrid orbitals to form the C—H and C—C σ-bonds. The remaining $2p_y$ and $2p_z$ orbitals are directed at 90° to one another and to the axis of the C—C bond so that, by lateral interaction, two π-bonds are formed.

In benzene there are thirty valency electrons. Each carbon atom contributes four, and each hydrogen one. It is known that the molecule is a plane regular hexagon, with angles of 120°, and it seems therefore that the carbon atoms must be in the trigonal (sp^2) state of hybridization, formed by the hybridization of $2s$, $2p_x$ and $2p_y$ orbitals. These hybrid orbitals, by overlapping with the hybrid orbitals of the neighbouring carbon atoms, form the C—C σ-bonds, as in fig. 1.7. The remaining hybrid orbitals overlap with the $1s$ orbitals of the six hydrogen atoms to form the C—H bonds. Twelve electrons are thereby allocated to the six single C—C bonds and another twelve to the six C—H bonds. This leaves six electrons which are not involved in the formation of the σ-bonds. These six electrons have been called the *aromatic sextet*

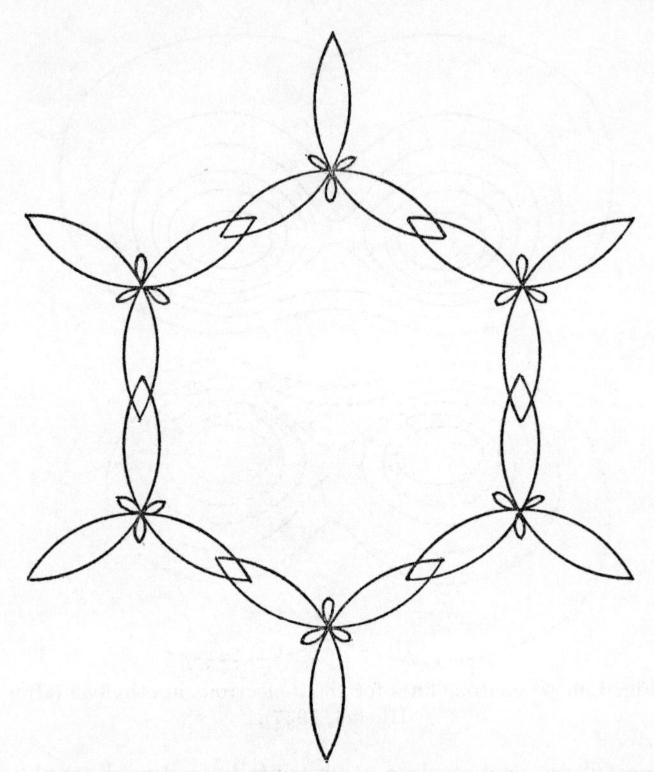

Fig. 1.7. Diagrammatic sketch showing the overlap of the trigonal hybrid orbitals (after Sklar, 1937).

(Armit & Robinson, 1925). By virtue of the trigonal state of hybridization, each carbon atom has a $2p_z$ orbital orientated perpendicular to the plane of the ring. These $2p_z$ orbitals interact laterally (as in ethylene) and because of the cyclic structure, with angles of 120°, these interactions must be the most perfect possible, forming very strong π-bonds.

From the six atomic orbitals of $2p_x$ character, six distinct molecular orbitals can be formed by the linear combination of atomic orbitals. These molecular orbitals, Ψ_1, Ψ_2, ..., Ψ_6, can each be represented by equations of the form

$$\Psi_1 = c_{1,1}\psi_1 + c_{1,2}\psi_2 + c_{1,3}\psi_3 + c_{1,4}\psi_4 + c_{1,5}\psi_5 + c_{1,6}\psi_6,$$

where ψ_1 represents the wave function which the electron would have if confined to nucleus 1, and ψ_2 the wave function which it would have if confined to nucleus 2, and so on. The constants,

$c_{1,1}$, $c_{1,2}$, ...; $c_{2,1}$, $c_{2,2}$, ...; etc., have various values, depending on the energy of the orbital. In the ground state, two electrons with antiparallel spins occupy the molecular orbital of lowest energy, and two occupy the molecular orbital of next lowest energy; the three molecular orbitals of highest energy content are unoccupied except in excited states.

In the lowest energy state the atomic orbitals interact to give a molecular orbital Ψ_1 which has a node in the plane of the ring. The resulting boundary surface therefore resembles two streamers which extend right around the ring, one above the plane, the other (of opposite phase) below the ring. In the other molecular orbitals there are additional nodal planes. The next two orbitals of higher energy have an additional nodal plane perpendicular to the plane of the ring. They are identical except that they are out of phase. The ground state of benzene may therefore be pictured as a superposition of the three molecular orbitals just described: two electrons in Ψ_1, two in Ψ_2 and two in Ψ_3. In the excited states, one or more electrons are promoted to a molecular orbital of higher energy content. The first three molecular orbitals are bonding orbitals, as their energies are less than that of an isolated $2p_z$ orbital. The other three are of higher energy and are antibonding.

Evaluation of the constants (Coulson, 1939) gives the following wave functions for the three occupied molecular orbitals:

$$\Psi_1 = \frac{1}{\sqrt{6}}(\psi_1+\psi_2+\psi_3+\psi_4+\psi_5+\psi_6),$$

$$\Psi_2 = \frac{1}{\sqrt{12}}(2\psi_1+\psi_2-\psi_3-2\psi_4-\psi_5+\psi_6),$$

$$\Psi_3 = \tfrac{1}{2}(\psi_2+\psi_3-\psi_5-\psi_6).$$

Using these values it is possible to derive a number of useful quantities. For example, the π-electron density at an atom 1 can be defined to be c_1^2, summed over all the occupied orbitals. At carbon atom 1, therefore, we have:

Orbital	Population (n)	c_1	nc_1^2
1	2	$1/\sqrt{6}$	1/3
2	2	$2/\sqrt{12}$	2/3
3	2	0	0
		π-electron density at	C_1 = 1·0

In the same way the π-electron density can be calculated for the remaining carbon atoms. It is 1·0 for each.

The above equations can also be used to derive the *bond order* for each carbon–carbon bond. The contribution of a π-electron to the mobile bond order, p, between neighbouring nuclei r and s, has been defined as $c_r c_s$. The total mobile bond order of the bond between r and s is therefore the sum of the contributions $c_r c_s$ from each of the six mobile electrons, i.e. $p = \Sigma c_r c_s$, and the total bond order $(1+p)$ is $1+\Sigma c_r c_s$. We have, therefore:

Orbital	Population of orbital (n)	c_1	c_2	nc_1c_2
1	2	$1/\sqrt{6}$	$1/\sqrt{6}$	1/3
2	2	$2/\sqrt{12}$	$1/\sqrt{12}$	1/3
3	2	0	1/2	0
			Mobile bond order	2/3

The mobile bond order is found to be 2/3 whichever pair of adjacent carbon atoms is chosen; the total bond order for each C—C bond is therefore 1·667.

The free-valence number of any carbon atom can then be derived from the bond orders. The free-valence number, F_r, at an atom r is defined as
$$F_r = N_{\max} - N_r,$$
where N_r is the sum of the orders of all bonds joining the atom r to the remainder of the system, and $N_{\max}$ is a constant depending on the chemical nature of the atom r and its state of hybridization. In aromatic systems the carbon atoms have a $N_{\max}$ value of 1·732. In benzene, for example, each C—C bond has a mobile order of 0·667, and each C—H bond a mobile order of 0. The free-valence number, F_r, for each carbon atom in benzene is, therefore, 1·732—(0·667 + 0·667 + 0), or 0·398.

Another approximate theoretical approach to the structure of benzene (and other aromatic compounds) has also been devised: this is based on the theory of resonance. It is assumed that the wave function can be expanded, or approximately represented, as a linear combination of the wave functions corresponding to all the possible valence-bond structures for a conjugated molecule. In other words, the actual normal state of the molecule is considered to be represented not by any one of the alternative reasonable structures of the classical valence-bond type, but by a combination, a hybrid, the individual contributions being determined by their nature and stability (Wheland, 1955).

A complete set of all the possible valence-bond structures which contribute to form a hybrid molecule is known as a canonical set. The hybrid molecule has a smaller energy content than any one of the contributory structures: the molecule may therefore be said to be stabilized by resonance. The canonical set for benzene (excluding structures in which there is charge separation) is (XIII *a*–*e*)

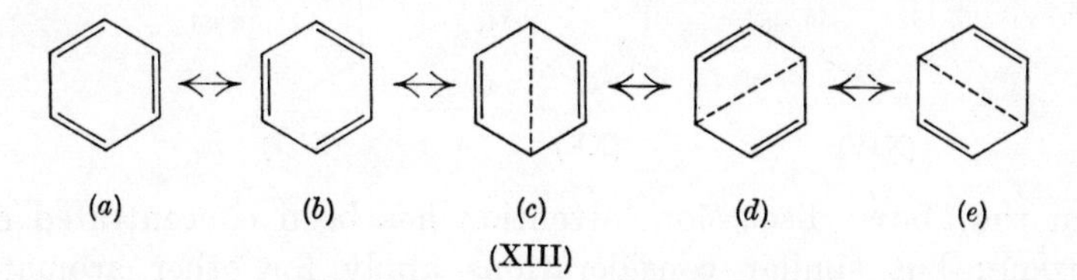

(XIII)

Benzene may be considered to be a hybrid of these five structures, or alternatively it may be said that the molecule resonates among the five structures, the symbol ↔ implying resonance. The Kekulé-type structures contribute to the hybrid to a greater degree than the Dewar-type structures, and the wave function for benzene can be represented by the expression:

$$\Psi_{\text{benzene}} = k_1(\psi_a+\psi_b)+k_2(\psi_c+\psi_d+\psi_e).$$

In this way it can be seen that in the hybrid benzene molecule all the carbon–carbon bonds are equivalent, and that they are neither single nor double, but are of intermediate character.

If the two Kekulé-type structures alone contribute to the benzene hybrid, then this hybrid can be represented by a regular hexagon having one π-electron along each side (that is, each carbon–carbon bond having 50 % double bond character). If the Dewar-type structures alone contribute to the hybrid, then this hybrid can be represented by a regular hexagon having 1/3 π-electron at each carbon atom and 2/3 π-electron along each bond. It has, however, been calculated that the Kekulé-type structures contribute 78 % to the hybrid and the Dewar-type structures 22 %. Using these percentages the distribution of π-electrons for the Kekulé-type structures and for the Dewar-type structures are given in (XIV) and (XV) respectively. The superposition of these two diagrams gives the 'molecular diagram' for benzene (XVI).

The figure 0·073, the 'concentration' of electrons at each carbon atom, has been called the free-valence number. The figure 0·927 represents the 'concentration' of electrons between each pair of adjacent carbon atoms. One half this value represents p, the mobile bond order, that is, the number of pairs of π-electrons between each two carbon atoms (Daudel & Daudel, 1948). The total bond order for the C—C bonds in benzene by this method is, therefore, 1·463.

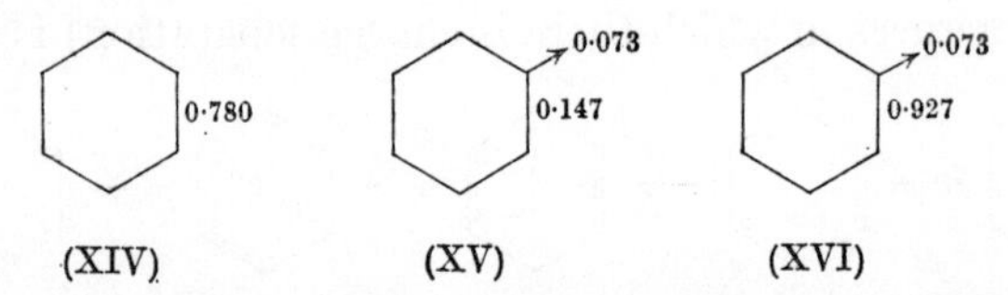

(XIV) (XV) (XVI)

In the above discussion, attention has been concentrated on benzene; but similar considerations apply for other aromatic compounds. Benzene belongs to a class of compound known as alternant hydrocarbons; that is, those compounds which possess only even-membered conjugated rings. In alternant hydrocarbons the carbon atoms can be divided into two sets, s (starred) and u (unstarred), such that each s-atom has only u-neighbours; and for such compounds the π-electron density at each carbon atom is 1·0. Non-alternant hydrocarbons are those conjugated hydrocarbons which contain odd-membered rings; and for these hydrocarbons (fulvene, azulene, etc.) the π-electron density at each carbon atom may be either greater than or less than unity.

1.5. Polycyclic benzenoid hydrocarbons. Benzene is the simplest and most important aromatic compound. Its reactions have been very extensively investigated, and the concept of aromatic character is largely based on its properties. More complex benzenoid hydrocarbons, such as naphthalene, anthracene, phenanthrene and pyrene, are however also of considerable interest, especially as they exhibit aromatic character in different degrees (Badger, 1954; Clar, 1964).

As with benzene, the reactions of naphthalene are predominantly those of substitution rather than of addition. Nitration, halogenation, sulphonation and other substitution reactions take place more readily than with benzene. With some polycyclic hydrocarbons, however, the reactions are predominantly those of addition and not of substitution. Such compounds are usually extremely reactive, and may be said to exhibit some aliphatic character.

TABLE 1.1. *Linear benzologues of benzene*

Compound	No. of linear rings	Colour of compound
Benzene	1	Colourless
Naphthalene	2	Colourless
Anthracene	3	Colourless
Naphthacene	4	Orange
Pentacene	5	Deep violet-blue
Hexacene	6	Deep black-green
Heptacene	7	Deep green-black

The series of linear benzologues of benzene is of special interest in this connection. This series (table 1.1) includes benzene, naphthalene, anthracene, naphthacene, pentacene, hexacene and heptacene. Each member is derived from the one immediately preceding it by the linear fusion of one additional benzene ring. The series is characterized not only by a deepening in the colour of the compounds from colourless to orange, deep violet-blue and deep green-black, but also by a very marked and progressive increase in reactivity. Pentacene is a deep violet-blue hydrocarbon of high reactivity; but heptacene is a deep green-black hydrocarbon which is so enormously reactive that it is impossible to obtain it in a pure state.

The series is also characterized by a marked increase in the tendency to give addition rather than substitution products as the number of rings increases; and this increase is paralleled by a marked and progressive increase in the stability of the dihydro derivatives. Although naphthalene is readily substituted by bromine and other reagents, anthracene (XVII) often reacts by addition under quite mild conditions, to give 9,10 addition products. With bromine, 9,10-dibromo-9,10-dihydroanthracene, or anthracene dibromide (XVIII) is formed. The reaction is reversible and, if the addition compound is treated with ice-cold hydriodic acid, or with substances such as phenol which take up bromine very readily, anthracene is regenerated. Anthracene dibromide slowly evolves hydrogen bromide, to give 9-bromoanthracene; but the reaction is not complete even after three days. Anthracene also reacts with nitric acid by addition, but the resulting compound can only be isolated as an ester. With nitrogen dioxide, anthracene gives 9,10-dinitro-9,10-dihydroanthracene; and with sodium, anthracene gives 9,10-disodio-9,10-dihydroanthracene.

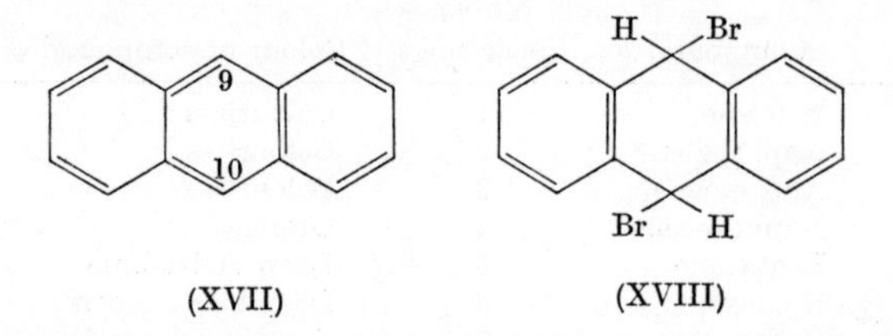

(XVII) (XVIII)

Other reagents give 9,10-*endo*-addition products. Maleic anhydride, for example, reacts with anthracene to give 9,10-dihydroanthracene-9,10-*endo*-$\alpha\beta$-succinic anhydride. The Diels–Alder reaction also occurs with a variety of other dienophiles, including *p*-benzoquinone, acrolein, allyl chloride, vinylacetate and acetylene dicarboxylic acid ester.

The photochemical addition of oxygen to anthracene and anthracene derivatives has also been widely studied. This reaction, which does not occur in the dark, leads to anthracene-9,10-peroxide (or photo-oxide). The nature of the solvent in this reaction is important, and photo-oxidations generally occur most readily in carbon disulphide; but tetramethylrubrene undergoes photo-oxidation even in the crystalline state.

The increasing stability of the dihydro-derivatives as the series is ascended is also noteworthy. Dihydrobenzene is relatively unstable, and 1,4-dihydronaphthalene is moderately stable, but 9,10-dihydroanthracene is a stable compound, and dihydro-derivatives of the higher members are also quite stable. Indeed, with some of the higher members, the tendency to form the dihydro-derivative is so marked that when the aromatic hydrocarbon is heated part is decomposed and the remainder converted into the dihydride. Hexacene, for example, is partly converted into dihydrohexacene during vacuum sublimation.

This progressive increase in the stability of the dihydrides also follows from studies of keto–enol tautomerism in the hydroxy-derivatives. As the series is ascended the hydroxy-derivatives become less important, and the corresponding keto-derivatives more important. Phenol reacts entirely in the enol form (XIX), and no derivatives of the keto-form (XX) can be prepared. Nevertheless, phloroglucinol shows many of the reactions of a triketone, and even resorcinol shows some of the reactions of a diketone. With 1-naphthol the keto–enol tautomerism again favours the enol form; but with 9-hydroxyanthracene the keto form (anthrone) is the more stable. Both tautomers can be isolated, and this substance reacts in either form depending on the conditions. With 5-hydroxy-naphthacene the keto–enol tautomerism favours the keto-form (XXII), and the enol form (XXI) is unimportant.

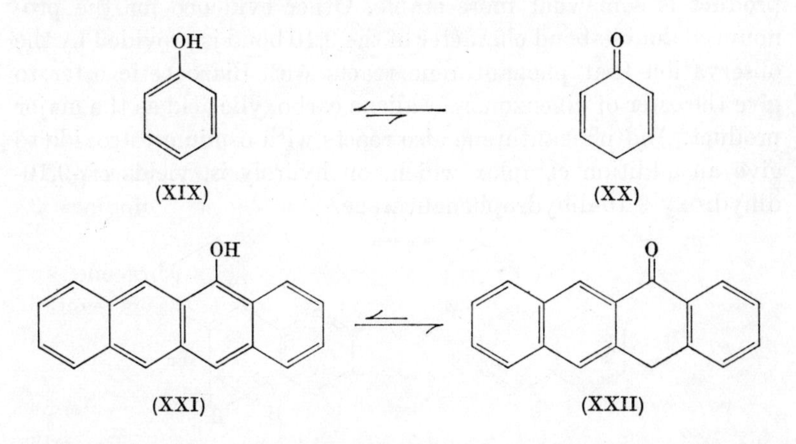

Similar *trans*-annular tautomerism has often been postulated for the methyl derivatives of this series of hydrocarbons. With the lower members (toluene, 1-methylnaphthalene and 9-methyl-anthracene) there is no doubt that the methyl derivatives are very much more stable than the methylene dihydro forms. On the other hand the spectroscopic evidence suggests that 5-methylpentacene (XXIII) exists almost entirely in the methylene dihydro form (XXIV) at ordinary temperatures (Clar & Wright, 1949).

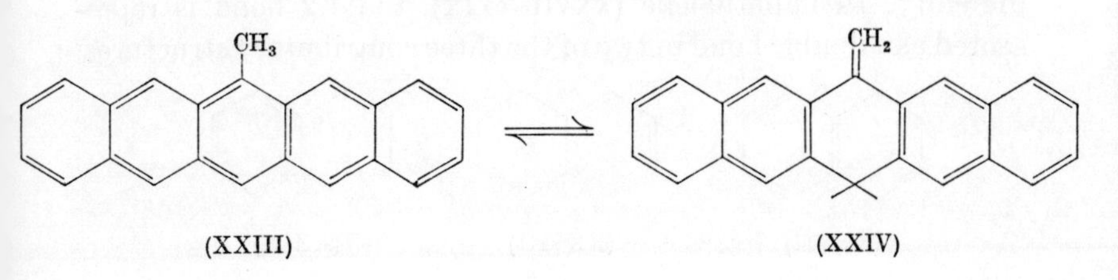

The angular fusion of benzene rings also gives compounds having varying degrees of aromatic character. Phenanthrene (XXV) is the simplest compound of this type, and is characterized by the pronounced reactivity and aliphatic character of the 9,10 bond. Phenanthrene is hydrogenated to the 9,10 dihydride over copper chromite catalyst at moderate temperatures. Bromine also adds to the 9,10 bond to give phenanthrene dibromide (XXVI); this loses hydrogen bromide on prolonged standing at room temperature, or more rapidly on heating, to give 9-bromophenanthrene. Chlorine also adds to phenanthrene in the same way, and this addition product is somewhat more stable. Other evidence for the pronounced double-bond character of the 9,10 bond is provided by the observation that phenanthrene reacts with diazoacetic ester to give the ester of dibenzonorcaradiene carboxylic acid as the major product. And phenanthrene also reacts with osmium tetroxide to give an addition complex which, on hydrolysis, yields *cis*-9,10-dihydroxy-9,10-dihydrophenanthrene.

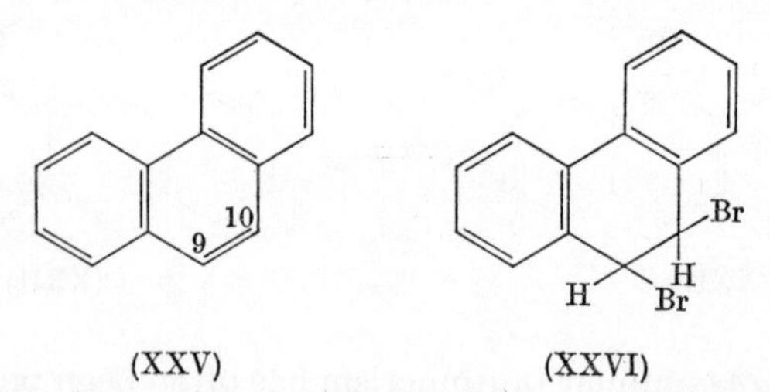

(XXV) (XXVI)

It is clear therefore that many polycyclic benzenoid hydrocarbons are much more reactive than benzene, and that many react with electrophilic reagents by addition rather than by substitution. These properties are reflected in the bond orders and free-valence numbers calculated for such hydrocarbons. Using simple resonance theory the double-bond characters of the carbon–carbon bonds in polycyclic benzenoid hydrocarbons can be evaluated without difficulty. In naphthalene (XXVII–XXIX), the 1,2 bond is represented as a double bond in two of the three contributory structures,

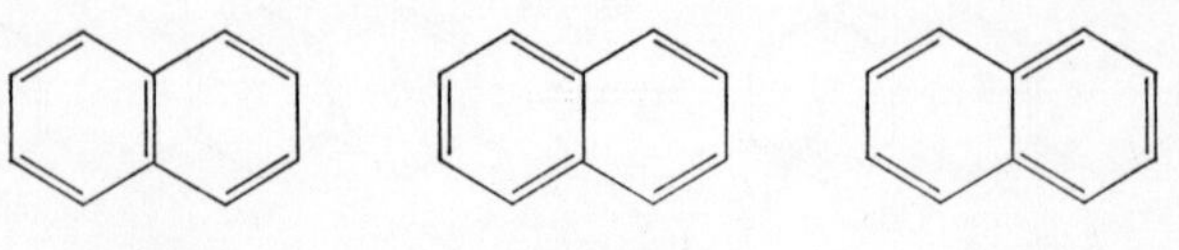

(XXVII) (XXVIII) (XXIX)

and may therefore be said to have $66\frac{2}{3}\%$ double-bond character. The 2,3 bond, on the other hand, is represented as a double bond in only one of the three structures, so it has $33\frac{1}{3}\%$ double-bond character. By appropriate summation the double-bond characters of all the bonds can be evaluated for naphthalene (XXX).

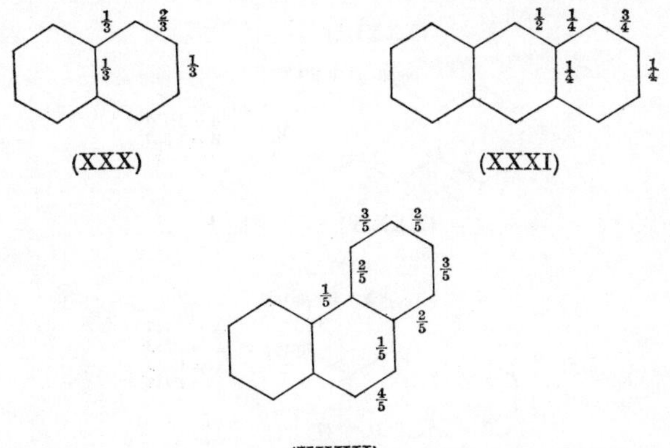

(XXX) (XXXI)

(XXXII)

The double-bond characters of the bonds in other polycyclic benzenoid hydrocarbons can be evaluated in the same way. For anthracene there are four Kekulé-type structures, so that the double-bond characters for this hydrocarbon (XXXI) can be calculated. Similarly, there are five Kekulé-type structures contributing to the resonance hybrid for phenanthrene, and this leads to the double-bond characters shown (XXXII).

The Daudel–Pullman method (see Pullman & Pullman, 1952) includes Dewar-type structures, and relatively complex molecular diagrams can be derived (XXXIII–XXXV). The figure adjacent to each carbon atom represents the free-valence number. The figure adjacent to each bond is the bond order. The bond order for the 1,2 bond in naphthalene is 1·527; and that for the 2,3 bond is 1·376.

It will be noted that the most reactive positions have relatively large free-valence numbers. In naphthalene, the 1-position has a larger free-valence number than the 2-position. In anthracene, the 9-position has a larger free-valence number than either the 1- or 2-position; and in phenanthrene, the 9- and 10-positions are also characterized by relatively large free-valence numbers. All these free-valence numbers are in accord with the known reactivities of the 1-position in naphthalene, the 9,10-positions in anthracene, and the 9,10-positions in phenanthrene.

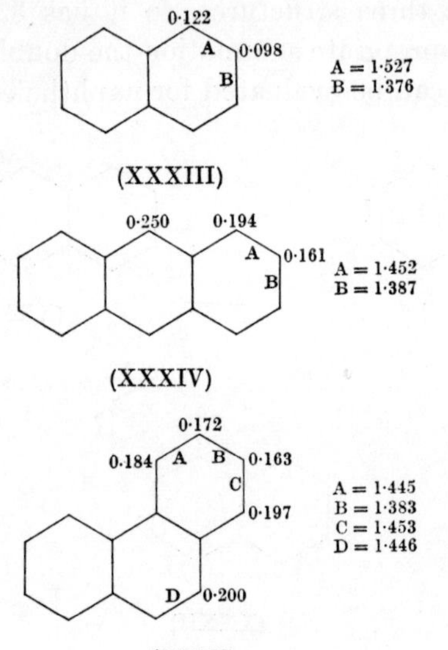

This method for the calculation of bond orders, based as it is on the valence-bond structures, becomes increasingly tedious as the molecules become more complex. The alternative method, based on the theory of molecular orbitals, is more attractive. It leads to bond orders and free-valence numbers which differ in magnitude from those derived by the valence-bond method, but which can also be related to chemical reactivity. This method, for benzene, has been described above (p. 16), and bond orders and free-valence numbers for naphthalene (XXXVI), anthracene (XXXVII), phenanthrene (XXXVIII) and pyrene (XXXIX) can be similarly calculated.

1.6. Heterocyclic aromatic compounds.

Six-membered compounds. Pyridine was first isolated in 1849 and it was not long before it was recognized that this substance is an analogue of benzene. The history of the various attempts to assign a satisfactory structural formula to pyridine closely parallels that relating to benzene.

There is no doubt that pyridine has aromatic character. It is even more resistant to oxidation than benzene. Its unsaturated nature is illustrated by the fact that it is readily reduced; but its reactions are predominantly substitution reactions. In this respect it is less reactive than benzene; it can be nitrated, but very severe conditions are required; and the same is true of sulphonation.

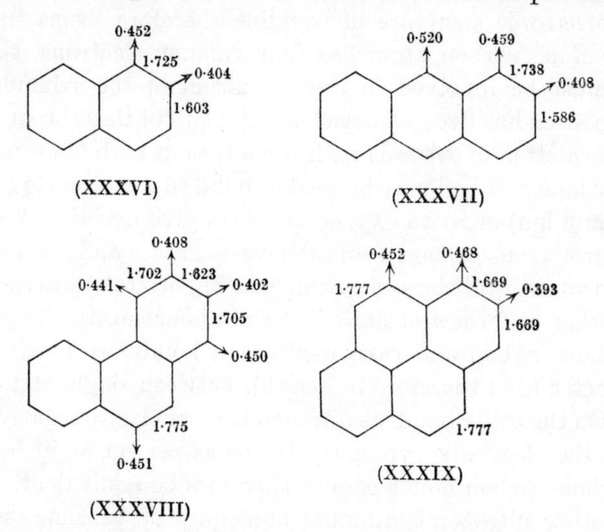

(XXXVI) (XXXVII) (XXXVIII) (XXXIX)

Although there are certain important differences to be considered, the solution to the pyridine problem closely resembles that already described for benzene. Unlike carbon, which has four valency electrons, nitrogen has five, and in most of its compounds it increases this number to form a stable octet. In ammonia, for example, the nitrogen atom is linked to three hydrogen atoms, and this compound is saturated with respect to this atom, leaving a 'lone pair' of electrons which are not involved in bonding. Ammonia is not saturated with respect to hydrogen ions, however, and the ammonium ion is formed by the addition of a proton. The nitrogen lone pair may also be used to form a linkage with an element, such as oxygen, which requires only two electrons to complete its octet.

The general shape and size of the pyridine molecule have been established experimentally. The carbon–carbon bonds have been found to be 1·39 Å, intermediate between the length characteristic of carbon–carbon single bonds (1·54 Å) and that of carbon–carbon double bonds (1·34 Å). Similarly, the carbon–nitrogen bonds have been found to be 1·34 Å, intermediate between the length characteristic of a carbon–nitrogen single bond (1·47 Å) and that of a carbon–nitrogen double bond (1·28 Å). The experimental determination of bond lengths in several pyridine derivatives confirms that these bonds are of intermediate character.

The electronic structure of pyridine therefore seems to be as follows. Each carbon atom has four valency electrons, three of which must be involved in the formation of the σ-bonds. The nitrogen atom has five valency electrons; two of these are involved in the formation of σ-bonds with the adjacent carbon atoms; two form the lone pair and may be used to bond to a proton (to give the pyridinium ion) or to an oxygen atom (to give pyridine *N*-oxide). Six electrons remain, one from each carbon atom and one from the nitrogen atom, and these six electrons ($2p_z$ orbitals) interact with one another to form π-orbitals just as in benzene. In this way all the carbon–carbon and carbon–nitrogen bonds are intermediate in character (and therefore in length), between single and double bonds. On the other hand, as nitrogen is more electronegative than carbon, the electronic symmetry is not as perfect as in benzene. The carbon–carbon bonds cannot therefore be equivalent.

The other nitrogen-containing analogues of benzene, such as pyridazine (XL), pyrimidine (XLI), pyrazine (XLII), 1,3,5-triazine (XLIII) and 1,2,4,5-tetrazine (XLIV) have similar electronic structures and must also be regarded as aromatic compounds. Polynuclear heterocyclic compounds can also be formed, and these bear the same relationship to pyridine and the other nitrogen-containing analogues of benzene as polynuclear aromatic hydrocarbons to benzene. Differences in properties compared with the corresponding hydrocarbons can be ascribed to the nitrogen atom.

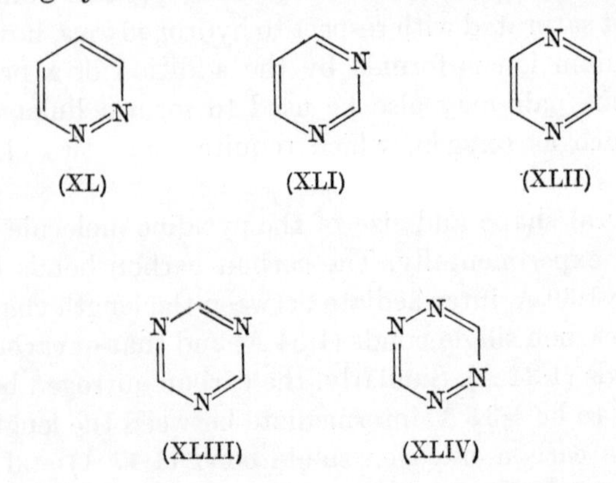

(XL) (XLI) (XLII)

(XLIII) (XLIV)

Heterocyclic compounds also exist in which the nitrogen occupies a bridge-head position. The nitrogen can be neutral, as in indolizine (XLV), which is a weak base, pk_a 3·94, or positively charged as in the quinolizinium ion (XLVI), or in the phenanthrene analogue (XLVII) having two quaternary nitrogen atoms.

(XLV)

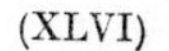

(XLVI)

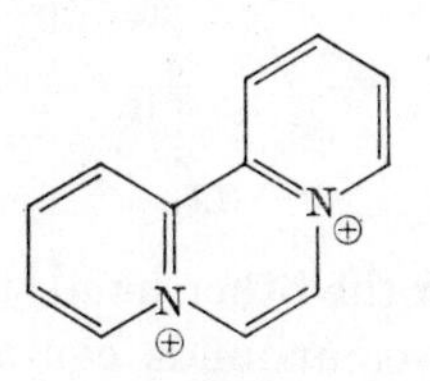

(XLVII)

In this connection, attention may be directed to the many heterocyclic compounds which have now been prepared and which contain a positively charged bridge-head nitrogen atom and a negatively charged bridge-head boron atom. The simplest example is 10,9-borazaronaphthalene (XLVIII) which is isoelectronic with naphthalene; but many other examples have also been prepared.

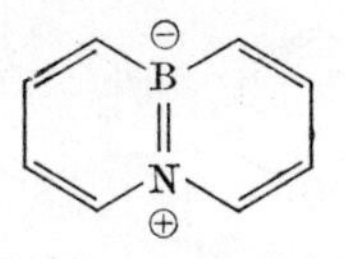

(XLVIII)

Heterocyclic compounds containing the groups >N—B<, >N—P< and >N—Si< can also be prepared. These groups are isostructural and isoelectronic with >C=C<; and phenanthrene analogues such as (XLIX), (L) and (LI) have been prepared.

Nitrogen is a member of Group V B of the Periodic Table, and six-membered heterocyclic compounds containing other members of Group V B (P, As, Sb, Bi) are theoretically possible; and a few

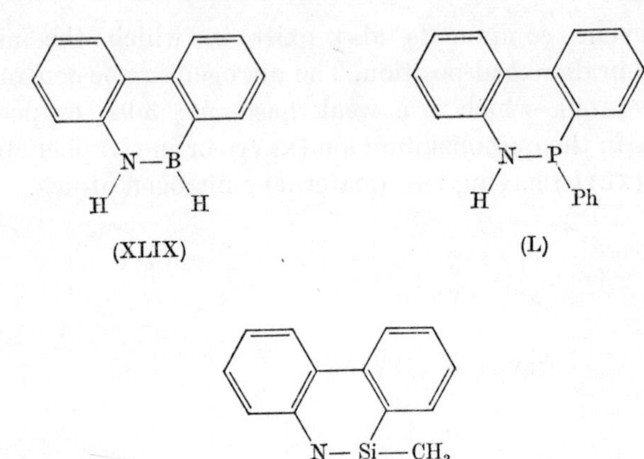

(XLIX) (L)

(LI)

have been prepared. On the other hand, neutral unsaturated six-membered heterocyclic compounds containing an element from Group VI B cannot be aromatic. These elements (O, S, Se, Te, Po) are divalent, and compounds such as α-pyran (LII) and α-thiapyran (LIII) contain only four π-electrons. An aromatic sextet can only occur if the ring carries a positive charge. The pyrylium cation (LIV) is such a system, and is aromatic: each carbon atom contributes one π-electron, and the oxygen atom likewise contributes one π-electron. The analogous thiapyrylium cation (LV) is also aromatic (Pettit, 1960).

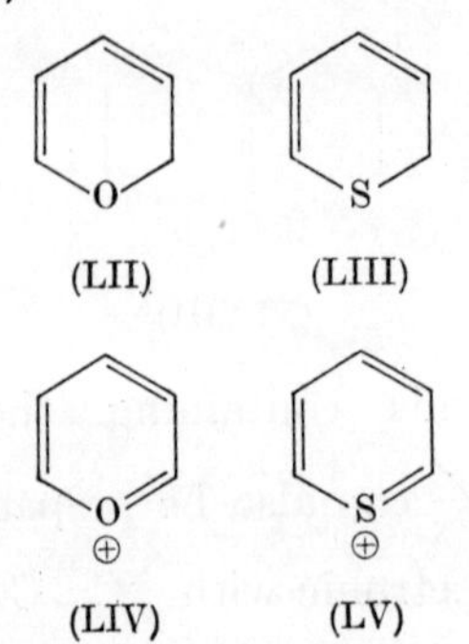

(LII) (LIII)

(LIV) (LV)

Pyrones and thiapyrones are derived from the parent systems by replacement of the methylene group by a carbonyl group. The resonance structures for these compounds indicate a contribution from a pyrylium or thiapyrylium system. With α-pyrone (LVI) the contribution from the pyrylium structures (LVI*b*, LVI*c*) must, however, be relatively small as this compound behaves essentially as an unsaturated lactone. It is readily hydrolyzed by alkali; and on catalytic hydrogenation it gives 5-valerolactone and valeric acid. It also behaves as a diene in the Diels–Alder reaction and readily adds maleic anhydride. On the other hand, γ-pyrone (LVII) behaves as an aromatic compound and the pyrylium structures (LVI*b* and LVI*c*) must therefore contribute significantly to the hybrid. The double bonds are not easy to reduce, and bromine reacts by substitution rather than by addition. Moreover the carbonyl group in γ-pyrones is relatively unreactive, and does not give the usual carbonyl derivatives. However, some γ-pyrones do react with activated methylene groups.

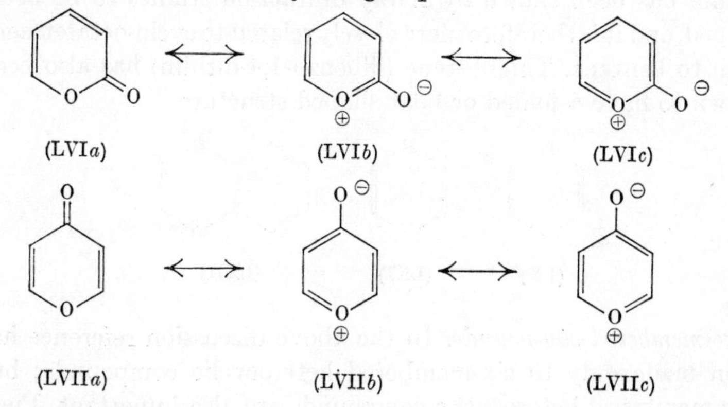

The situation is not unlike that in the 2- and 4-hydroxypyridines. These hydroxy derivatives exhibit lactim–lactam tautomerism, and both tautomers are aromatic. 4-Hydroxypyridine (LVIII), for example, is aromatic, and is tautomeric with 4-pyridone (LIX) which must also be regarded as an aromatic compound.

Heterocyclic compounds containing two hetero-atoms from Group VI B are not aromatic. 1,4-Dioxin (LX), for example, behaves as an unsaturated aliphatic ether, comparable to vinyl ether. With 1,4-dithiin (LXI) there is a greater tendency towards cyclic conjugation in accordance with the known ability of sulphur to transmit conjugation. Benzo-1,4-dithiin (LXII) undergoes a number of electrophilic substitution reactions. It is nitrated in the 2-position, it is acetylated by acetic anhydride, and with *N*-methyl-

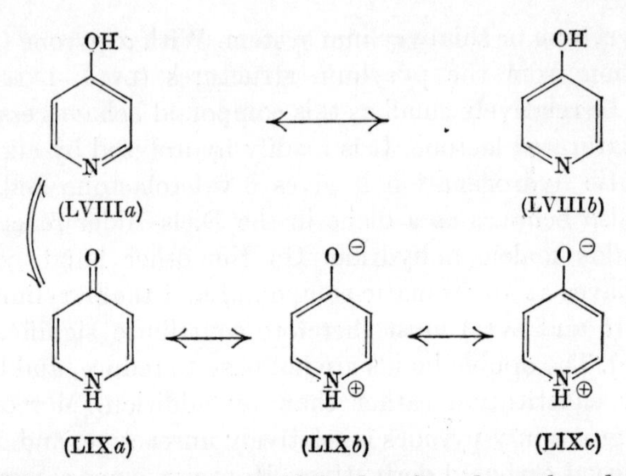

formanilide it gives the 2-aldehyde. On the other hand, 1,4-dithiin has been shown by X-ray diffraction studies to be boat-shaped, and it is therefore more closely related to cyclo-octatetraene than to benzene. Thianthrene (dibenzo-1,4-dithiin) has also been shown to have a folded or boat-shaped structure.

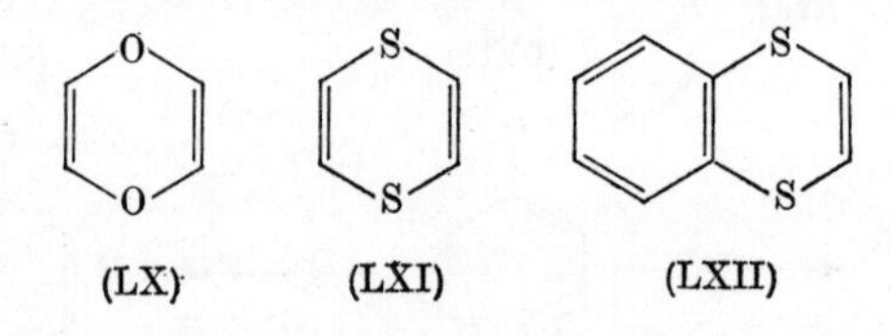

Five-membered compounds. In the above discussion reference has been made only to six-membered heterocyclic compounds; but five-membered heterocyclic compounds are also important. Five-membered heterocyclic compounds containing two formal double bonds in an alternating sequence with the hetero-atom must be regarded as aromatic compounds. Thiophen, furan and pyrrole are well-known examples of such systems; but analogous ring systems containing Se, Te, P, Ge, Sn, As and Sb have all been prepared. In addition, many related compounds are known which contain more than one hetero-atom.

Analysis of the infrared, Raman, and microwave spectra of pyrrole indicates that the atoms are, almost certainly, all coplanar (Wilcox & Goldstein, 1952; Kofod, Sutton & Jackson, 1952). The dimensions of the molecule have been determined by microwave spectroscopy: the 2,3 and 4,5 bonds are 1·37 Å in length, the 3,4 bond is 1·42 Å, and the 1,2 bond is 1·38 Å (Bak *et al.* 1956).

All the carbon–carbon and carbon–nitrogen bonds are therefore intermediate in length between those characteristic of pure single bonds and pure double bonds, and although the bond angles in a five-membered ring must differ significantly from 120°, the electronic structure for pyrrole can be inferred. The valency electrons of the four carbon atoms must be hybridized in an approximately trigonal fashion; and the valency electrons of the nitrogen atom must also be trigonally hybridized. Two of the nitrogen orbitals overlap with the trigonally hybridized orbitals of the neighbouring carbon atoms to form carbon–nitrogen σ-bonds, and one overlaps with the $1s$ orbital of a hydrogen atom to form the N—H bond. The remaining two electrons occupy an orbital, which is perpendicular to the plane of the ring, and which overlaps laterally with the $2p_z$ orbitals provided by the carbon atoms. The resulting molecular orbitals form streamers above and below the ring, just as in benzene. However, as the bond angles are all less than the 120° required for perfect sp^2 hybridization, the conjugation in pyrrole must be less perfect than in benzene. Moreover, nitrogen is more electronegative than carbon, and the nitrogen-orbital perpendicular to the ring may be thought of as contracted relative to the carbon $2p_z$ orbitals. The π-electron densities for the different atoms in pyrrole can be calculated, but the actual figures depend on the assumptions made.

The nitrogen atom in pyrrole, unlike that in ammonia or in pyridine, has no lone pair of electrons not involved in bond formation. Pyrrole can only form a salt (and it does so by adding a proton to one of its carbon atoms) at the expense of its aromaticity, a fact which explains not only its weakly basic nature, but also its tendency to resinify when treated with strong acids. Imidazole (LXIII) is also of interest in this connection as it is a mono-acidic base. This is reasonable if the nitrogen atom at the 1-position contributes its lone pair to the common pool of π-electrons (as in pyrrole), and the nitrogen atom at the 3-position contributes only one electron, leaving its lone pair available for salt formation, as in (LXIV).

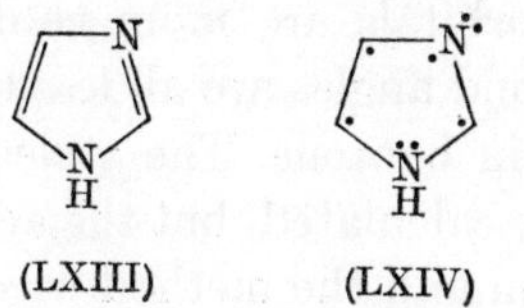

(LXIII) (LXIV)

In an alternative approach pyrrole may be considered as a resonance hybrid of a number of contributing structures (LXV). This method has the advantage of emphasizing the result of the delocalization of the lone pair, namely that the carbon atoms of the ring acquire some negative charge so that pyrrole is very reactive towards electrophilic reagents.

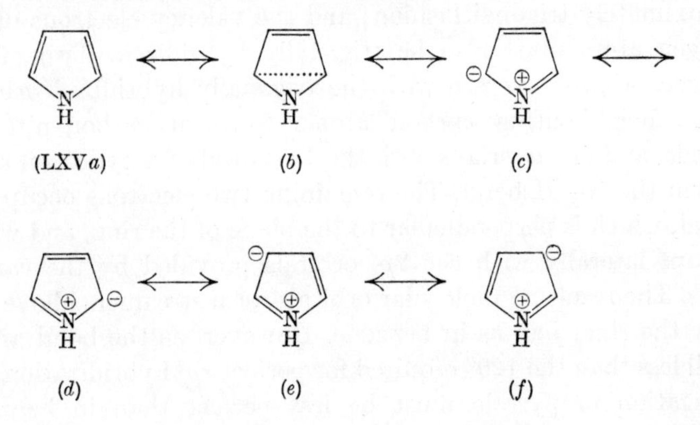

Evidence for cyclic conjugation in furan is also provided by the experimental determination of bond lengths. The 2,3 bond in furan has been found to be about 1·35 Å, and the 3,4 bond about 1·44 Å; and similar lengths were found for these bonds in 2-furoic acid. In other words the carbon–carbon bonds are intermediate in length between that characteristic of a single bond and that of a double bond; but it must be admitted that the length of the carbon–oxygen bond in furan is close to that of a carbon–oxygen single bond.

The electronic structure of furan must be similar to that of pyrrole. Each carbon atom is in an approximately trigonal state of hybridization with a $2p_z$ orbital perpendicular to the plane of the ring. These orbitals interact laterally with the two $2p$ electrons provided by the oxygen atom, and the resulting molecular orbitals again form two streamers above and below the ring, just as in pyrrole. However, as oxygen is much more electronegative than carbon the oxygen orbitals are more contracted than those of carbon, and as the bond angles are all less than 120°, conjugation is less efficient than in benzene. The π-electron densities for the atoms in furan can be calculated, but the actual figures depend on the parameters chosen and the method used.

The alternative theoretical approach to the structure of furan also provides a satisfactory explanation for the known facts. Furan may be regarded as a resonance hybrid of a number of contributing structures (LXVI), and this leads to the conclusion that the carbon–carbon bonds are of intermediate double-bond character and that the carbon atoms acquire some negative charge.

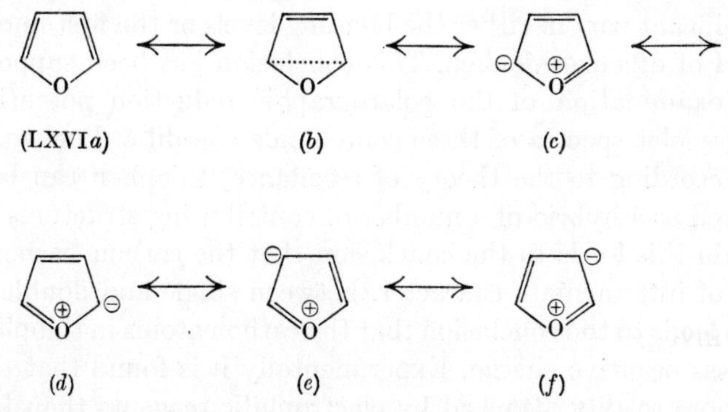

Finally, attention may be directed to thiophen. Several experimental studies (p. 39) indicate that the C—C bonds in thiophen are intermediate in length between that characteristic of a single bond and that of a double bond.

The electronic structure of thiophen must be very similar to that of benzene, pyrrole and furan. Each carbon atom must be in an approximately trigonal state of hybridization, with a $2p_z$ orbital perpendicular to the plane of the ring. These orbitals interact laterally with the two $3p$ electrons provided by the sulphur atom, forming π-bonds, as in pyrrole and furan.

Alternatively, it has been suggested that as the energy difference between $3p$ and $3d$ orbitals is not great, pd hybridization of the sulphur orbitals could occur to give pd^2 hybrid orbitals two of which have the correct symmetry and energy to conjugate with the carbon $2p_z$ orbitals (Longuet-Higgins, 1949; de Heer, 1954). The third hybrid is high in energy and would be unoccupied in the ground state. This suggestion has been examined experimentally (Gerdil & Lucken, 1965). The electron spin resonance spectra of the radical anions of dibenzothiophen, dibenzofuran and dibenzoselenophen were determined and the hyperfine splitting constant assigned to the various protons. The hyperfine splitting constants for dibenzothiophen were compared with those calculated using two models for conjugation by the sulphur atom: with and without participation of the sulphur 3*d* orbitals. It was found that the *d*-orbital model predicts an incorrect odd-electron distribution in the radical anion, but that good agreement with experiment was afforded by a *p*-orbital model for both dibenzofuran and dibenzothiophen. It was therefore concluded that 3*d* orbitals play no significant part in either the bonding levels or the first unoccupied level of dibenzothiophen. This conclusion has been supported by an examination of the polarographic reduction potentials and ultraviolet spectra of these compounds (Gerdil & Lucken, 1966).

According to the theory of resonance, thiophen can be represented as a hybrid of a number of contributing structures (LXVII). Again this leads to the conclusion that the carbon–carbon bonds are of intermediate character, between single and double; and it also leads to the conclusion that the carbon atoms in thiophen have excess negative charge. Experimentally it is found that thiophen is more readily attacked by electrophilic reagents than benzene; but it is less reactive than furan. If 3*d* orbitals of the sulphur atom are involved, additional resonance structures must be included.

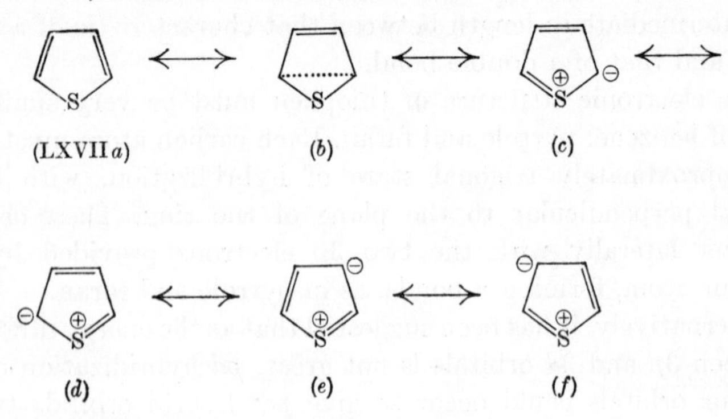

Sydnones. When *N*-phenylglycine is treated with nitrous acid and the resulting *N*-nitroso-*N*-phenylglycine (LXVIII) is dehydrated with acetic anhydride, 3-phenylsydnone is formed. The reaction is a general one for aryl- and alkyl-glycines, and was discovered in Sydney in 1935. No single satisfactory classical valence-bond structure can be drawn for the sydnone ring system, and the apparent anomaly of valency attracted considerable attention in the years following the discovery of these compounds (Baker &

Ollis, 1957). However, it was soon recognized that the sydnones are aromatic compounds: seven electrons are provided by the annular atoms, but if one is transferred to the carbonyl oxygen a sextet is obtained. 3-Phenylsydnone may therefore be represented as a resonance hybrid of a number of polar structures, such as (LXIX), or it can be more readily represented by (LXX).

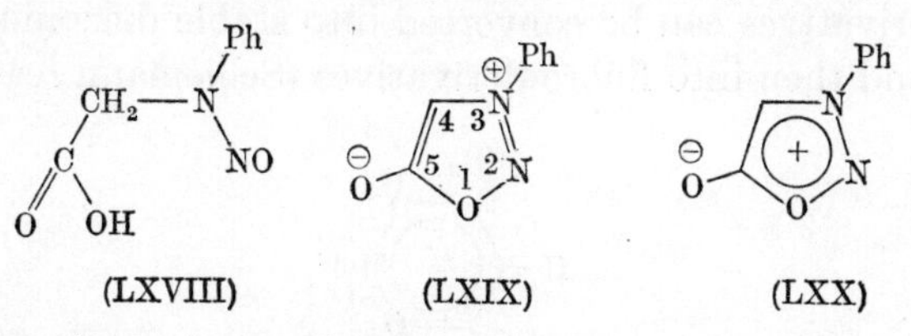

(LXVIII) (LXIX) (LXX)

Sydnones have large dipole moments, *ca.* 5–7*D*; and their aromatic nature is confirmed by the fact that they undergo ready substitution. 3-Phenylsydnone, for example, is substituted by electrophilic reagents at the 4-position, and its reactivity has been compared with that of thiophen. It is also of interest that 3-methyl-4-phenylsydnone (LXXI) is readily nitrated in the phenyl ring, suggesting that structure (LXXI*b*) contributes significantly to the resonance hybrid.

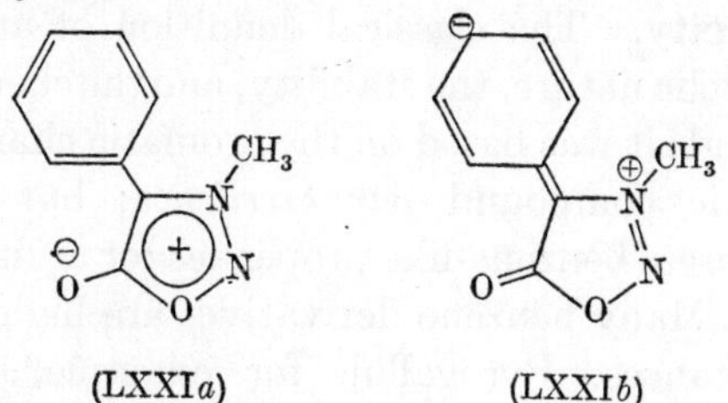

(LXXI*a*) (LXXI*b*)

Several ring systems analogous to the sydnones have also been prepared. Examples are the compound (LXXII) which incorporates an oxazole skeleton, and the 1,3,4-thiadiazole derivative (LXXIII).

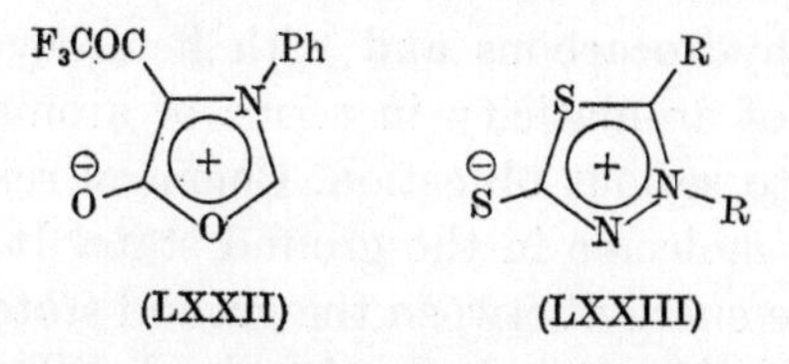

(LXXII) (LXXIII)

Organo-metallic compounds. Metal acetylacetonates (LXXIV) were first postulated to be aromatic systems in 1925 (Armit & Robinson, 1925), but the chemistry of these heterocyclic compounds has been studied only in more recent years. Metal acetylacetonates are substituted by dinitrogen tetroxide, and by nitrosyl chloride; and substitution with electrophilic reagents has been found to be quite general (Collman, 1965). Such substitutions include Friedel–Crafts acylation, and Mannich-type condensations. Moreover, the aromatic nature of such compounds is illustrated by the fact that amino derivatives can be converted into stable diazonium fluoroborates and then into fluoro-derivatives (Schiemann reaction).

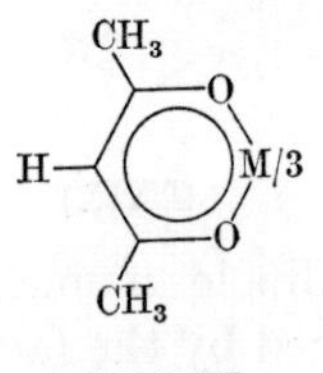

(LXXIV)

In compounds of this type it seems possible that each chelate ring has a cyclic π-orbital formed by overlap of vacant *d*-orbitals of the central metal atom with the six π-electron orbitals of the ligand.

1.7. Aromaticity. The classical definition of aromaticity was based on the cyclic nature, the stability, and the chemical reactivity of the compound: it was based on the aromatic character. Benzene is the aromatic compound *par excellence*; but relatively few compounds possess benzene-like properties (or aromatic character) in high degree. Many benzene derivatives are highly reactive and unstable substances. Pyrogallol, for example, is very easily oxidized, even by air; many polycyclic benzenoid compounds are much less stable and much more reactive than benzene; and many polycyclic benzenoid compounds undergo addition reactions rather than substitution reactions. Similar difficulties arise with non-benzenoid hydrocarbons and with heterocyclic compounds. Any definition of aromaticity in terms of aromatic character is therefore open to serious objection. Chemical reactivity is not a property of the molecule in the ground state. It depends on the difference in free energy between the ground state and the transition state for the chemical change involved. When this difference is small the compound is reactive irrespective of the energy content in the ground state (Craig, 1959*a*; Sondheimer, 1963).

Although aromaticity is not a function of the stability and chemical reactivity, it is a function of the electronic structure, and it is the main object of the present essay to define aromaticity with reference to the electronic structure of the molecule or ion in question. *An unsaturated cyclic or polycyclic molecule or ion (or part of a molecule or ion) may be classified as aromatic if all the annular atoms participate in a conjugated system such that, in the ground state, all the π-electrons (which are derived from atomic orbitals having axial orientation to the ring) are accommodated in bonding molecular orbitals in a closed (annular) shell.*[1]

This definition of an aromatic molecule or ion has several important consequences. Some of these can be studied experimentally, and it is therefore possible to determine whether or not a given molecule or ion does in fact have the electronic structure required for aromaticity.

[1] I am grateful to Dr G. E. Lewis and Dr J. A. Elix for helpful discussions on this definition.

2. Some consequences of aromaticity

2.1. Bond lengths. The length of a carbon–carbon single bond, such as occurs in diamond and in the paraffins, is near 1·54 Å. The length of a carbon–carbon double bond, as in ethylene, is about 1·34 Å. In all non-conjugated molecules these bond lengths, with only minor variations, are observed; but experiment has shown that in conjugated compounds there are considerable deviations from these bond lengths. In conjugated molecules the bonds which, in the classical valence-bond structure, are represented as single bonds are found to be significantly shorter than 1·54 Å; and those which are represented as double bonds are longer than 1·34 Å.

With conjugated polyenes, and with conjugated non-aromatic cyclic polyenes, the carbon–carbon bonds are alternately long and short, as required by the valence-bond structure. With aromatic compounds, however, some variation in bond length may occur; but, in general, bond-length alternation is not observed.

With benzene, conclusive experimental evidence has been obtained that the bonds do not alternate in length (as would be required by the Kekulé formula), but that they are all equivalent and of length 1·39 Å (see chapter 1). With naphthalene and with the polycyclic benzenoid hydrocarbons some bond-length variation occurs, the shortest bonds being those which are calculated to have the largest bond orders (Robertson, 1948; Trotter, 1964). In naphthalene (LXXV) the 1,2-bond (indicated by thick lines) is significantly shorter than the 2,3, the 1,9, and the 9,10 bonds, and also significantly shorter than the carbon–carbon bonds in benzene. Similarly in anthracene (LXXVI), in tetracene (LXXVII), and in pentacene (LXXVIII), the 1,2-bonds are especially short.

Bond-length variations have also been observed with the highly condensed benzenoid hydrocarbons such as pyrene (LXXIX) and coronene (LXXX). With both these hydrocarbons the shortest bonds (indicated by thick lines) are on the periphery.

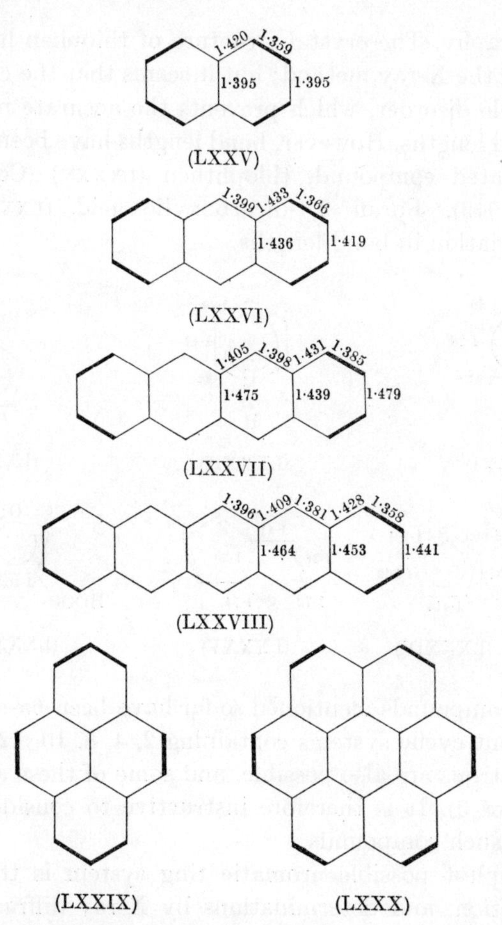

(LXXV)

(LXXVI)

(LXXVII)

(LXXVIII)

(LXXIX) (LXXX)

Bond lengths have been determined in many heterocyclic aromatic compounds. It is again found that the carbon–carbon bonds are of intermediate length, and that some bond-length variation occurs. In pyridine (LXXXI), the carbon–carbon bonds are all of length 1·39 Å, and the carbon–nitrogen bonds of length 1·37 Å; and in pyridine hydrochloride (LXXXII), the carbon–carbon bonds are 1·40 and 1·42 Å.

Several five-membered heterocyclic compounds have been examined and bond-length variations, in general accord with calculated bond orders, have been observed. The bond lengths in thiophen (LXXXIII) have been determined by microwave spectroscopy, and in thiophen-2-carboxylic acid (LXXXIV) by X-ray crystallography. The crystal structure of thiophen has also been studied by the X-ray method; but it seems that the crystals show considerable disorder, which prevents the accurate measurement of the bond lengths. However, bond lengths have been determined in the related compound, thiophthen (LXXXV) (Cox, Gillot & Jeffrey, 1949). Furan 3,4-dicarboxylic acid (LXXXVI) shows similar variation in bond lengths.

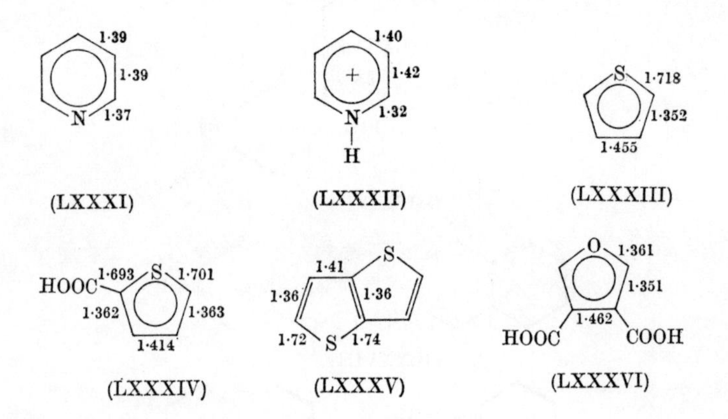

(LXXXI) (LXXXII) (LXXXIII)

(LXXXIV) (LXXXV) (LXXXVI)

All the compounds mentioned so far have been 6π-electron ring systems; but cyclic systems containing 2, 4, 8, 10, 12, 14, 16, 18, etc., π-electrons are also possible, and some of these are aromatic (see chapter 3). It is therefore instructive to consider the bond lengths in such compounds.

The simplest possible aromatic ring system is the cyclopropenium cation, and determinations by X-ray diffraction of the crystal structure of triphenylcyclopropenium perchlorate (LXXXVII) show that all the carbon–carbon bonds in the three-membered ring are equivalent, and of length $1{\cdot}373 \pm 0{\cdot}005$ Å (Sundaralingam & Jensen, 1963, 1966). On the other hand, the carbon–carbon bonds in the non-aromatic cyclo-octatetraene (LXXXVIII) alternate in length as required by the valence-bond structure. According to electron diffraction experiments (Karle, 1952) the carbon–carbon single bonds are of length 1·50 Å, and the double bonds 1·35 Å. X-ray crystallographic studies give 1·54 Å for the single bonds and 1·34 Å for the double bonds (Kaufman, Fankuchen & Mark, 1947, 1948). Similarly, X-ray analysis of cyclo-octatetraene carboxylic acid gives 1·470 Å for the single bonds, and 1·322 Å for the double bonds; and octaphenylcyclo-octatetraene also exhibits bond alternation with average carbon–carbon bond lengths of 1·493 and 1·343 Å.

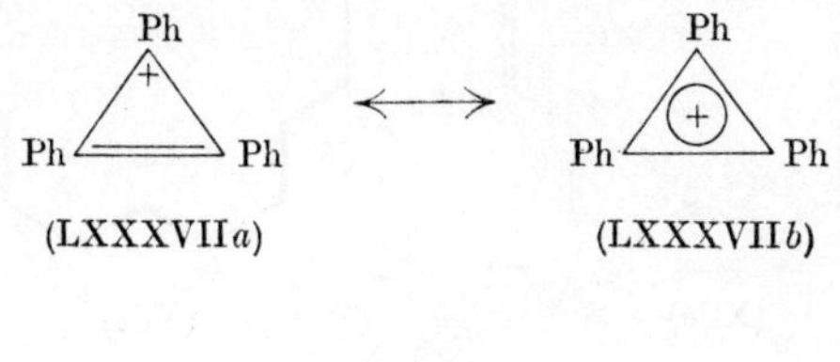

(LXXXVII*a*) (LXXXVII*b*)

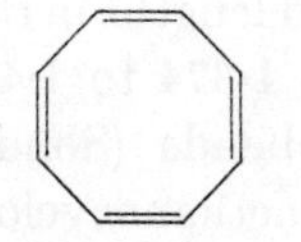
(LXXXVIII)

Seven-membered ring systems having six π-electrons and a positive charge can also display aromaticity. The cycloheptatrienium cation itself is the simplest of these; but several derivatives are known (see chapter 3). Among the most important of these is tropolone. The X-ray crystallographic study of cupric tropolone (LXXXIX) shows this ring system to be an almost regular, planar heptagon with average C—C bond lengths of 1·40 Å; and this has been confirmed by studies on sodium tropolonate, and on tropolone hydrochloride.

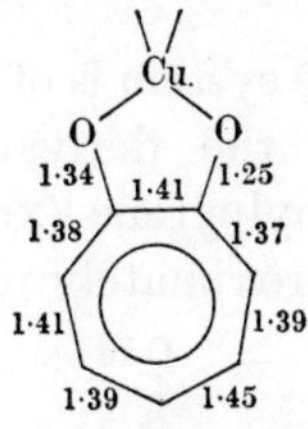

(LXXXIX)

Interesting results have been obtained with the annulenes (see chapter 3). [18]Annulene (XC) is undoubtedly aromatic, despite its relative instability. It has been found that the molecules are situated at crystallographic centres of symmetry, so a structure with alternate long and short C—C bonds is ruled out (Sondheimer, 1963; Bregman *et al.* 1965). In fact, the C—C bonds have been shown to be either 1·419 Å (indicated by thin lines), or 1·382 Å (indicated by thick lines).

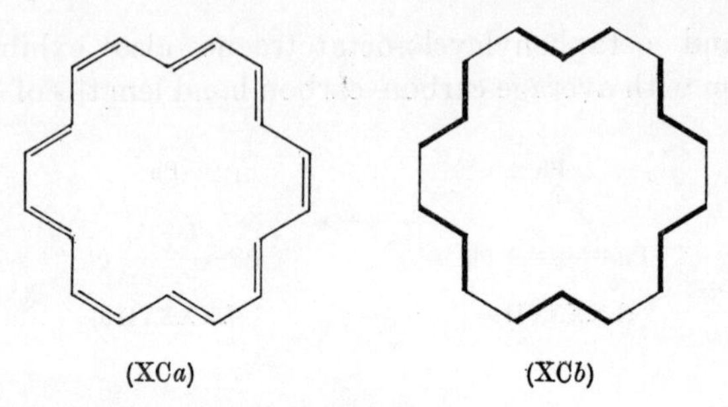

(XC*a*) (XC*b*)

The carbon–carbon bond lengths in the aromatic bisdehydro[14]-annulene (XCI) vary from 1·374 to 1·400 Å, with 1·200 Å for the bonds written as triple bonds (Sondheimer, 1963). An X-ray structure analysis of 1,6-methanocyclodecapentaene 2-carboxylic acid (XCII), which has an aromatic ring system, has shown that this compound has an almost planar perimeter with no appreciable bond alternation (Dobler & Dunitz, 1965).

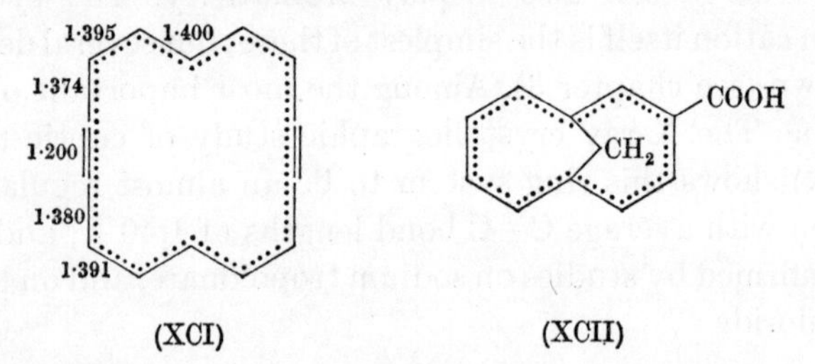

(XCI) (XCII)

The 15,16-dihydropyrene system is of special interest (see p. 98). The X-ray structure of the derivative, 2,7-diacetoxy-15,16-dihydro-*trans*-15,16-dimethylpyrene (XCIII), has shown that this compound also has an approximately planar perimeter, in accord

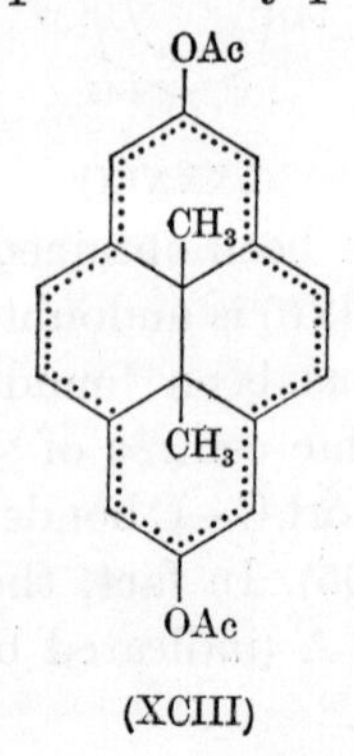

(XCIII)

with its undoubted aromaticity. The average carbon–carbon bond length is 1·395 Å, with only small variations (Hanson, 1965).

Finally, reference may be made to the sandwich structure of the organo-metallic, aromatic compound, ferrocene (XCIV) (see chapter 4). In this compound the C—C bond lengths have been found to be 1·403 ± 0·02 Å (Dunitz, Orgel & Rich, 1956).

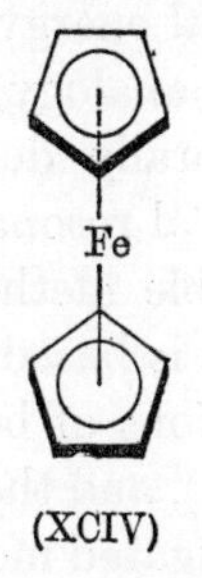

(XCIV)

It can be concluded that the average carbon–carbon bond length in aromatic compounds is near 1·39 Å, but that considerable bond-length variation occurs. This variation seems to be related to the bond orders of the various carbon–carbon bonds, and not to the single and double bonds in any one classical valence-bond structure. Bond-length alternation does not, in general, occur in aromatic compounds. Indeed, the observed bond-length alternation in cyclo-octatetraene is *prima facie* evidence that this compound does not exhibit aromaticity. By the same token, the absence of bond-length alternation in [18]annulene provides strong support for its aromaticity.

2.2. Resonance energies. In the previous chapter it was pointed out that the π-electrons of conjugated molecules occupy molecular orbitals which embrace all the carbon atoms in the conjugated system. These π-electrons have a greater bonding energy than they would have if paired in localized bonds, and this increase in bonding energy is of the utmost importance. For polyenes and other non-aromatic conjugated molecules the increase in energy is small; but the cyclic delocalization which occurs in aromatic molecules or ions gives rise to relatively large increases in bonding energy. In molecular orbital theory the delocalization energy is the calculated additional bonding energy which results from the delocalization of the electrons originally constrained in isolated double bonds. In valence-bond theory the vertical resonance energy is the difference between the calculated energy of the most stable single contributing structure which has the same geometry as the molecule in question (i.e. the structure of lowest energy), and the true energy of the molecule.

The calculation of vertical resonance energies from first principles is not possible; but empirical energy values for individual structures can be determined by analogy with other compounds; and comparison with experimentally determined energies of actual molecules then gives empirical resonance energies.

The most widely applicable method for the determination of empirical resonance energies is based on the heats of combustion. Tables of average contributions of bonds to heats of combustion have been devised (table 2.1), and the calculated value of the heat of combustion of a non-conjugated molecule, or of the contributing structure of lowest energy for a conjugated molecule, can therefore be obtained. For conjugated molecules the actual heat of combustion is lower than that calculated for the contributing structure; the difference is the empirical resonance energy. For example, the calculated heat of combustion for Kekulé benzene is the sum of the heats of combustion of three C—C bonds, three C═C bonds, and six C—H bonds, and the correction for a six-membered ring, or 825·1 kcal mole^{-1}. The observed heat of combustion of benzene is 789·1 kcal mole^{-1}, so the empirical resonance energy of benzene is 36·0 kcal mole^{-1}. Empirical resonance energies determined from heats of combustion data are subject to large probable errors, being derived from the difference between two relatively large values, the difference itself being relatively small. Nevertheless heats of combustion have been determined for very many compounds, and some values have been collected in table 2.2.

There is an alternative, and more accurate method for the determination of empirical resonance energies; but it is not applicable to so wide a range of compounds. This method (Conant & Kistiakowsky, 1937; Wheland, 1955) is based on experimentally determined heats of hydrogenation. The heat of hydrogenation of ethylene is 32·8 kcal mole^{-1}; that of compounds of type $CH_2{=}CH.CH_2R$ averages 30·2 kcal mole^{-1}; and that of compounds of type $RCH_2CH{=}CH.CH_2R'$ averages 28·2 kcal mole^{-1}, R being H or alkyl. These variations are due to the effects of the alkyl

TABLE 2.1. *Some bond contributions to heats of combustion*
(Wheland, 1955)

Bond	Contribution kcal mole⁻¹	Remarks
C—H	54·0	
C—C	49·3	
C═C	121·2	Ethylene
—	119·1	Monosubstituted ethylenes
—	117·4	*cis*-1,2-Disubstituted ethylenes, including six-membered rings
—	115·7	*cis*-1,2-Disubstituted ethylenes in five-membered rings
N—H	30·5	
C—N	33·0	
—	+1·0	Correction for six-membered ring
—	+6·0	Correction for five-membered ring

TABLE 2.2. *Heats of combustion and empirical resonance energies*
(Wheland, 1955)

Substance	Heat of combustion (kcal mole⁻¹) Observed	Calculated	Empirical resonance energy (kcal mole⁻¹)
Ethane	372·820	373·3	—
Ethylene	337·234	337·2	—
cis-2-Butene	648·115	648·0	—
1,3-Butadiene	608·5	611·5	3·0
1,3-Cyclopentadiene	707·7	709·3	1·6
2,3-Dimethyl-1,3-butadiene	917·3	918·9	1·6
Benzene	789·1	825·1	36·0
Naphthalene	1249·7	1310·7	61·0
Anthracene	1712·1	1795·6	83·5
Phenanthrene	1705·0	1796·3	91·3
Naphthacene	2170·6	2280·6	110·0
Benz[*a*]anthracene	2169·8	2281·4	111·6
Benzo[*c*]phenanthrene	2172·5	2282·1	109·6
Furan	506·9	522·7	15·8
Thiophen	612·0	640·7	28·7
Pyrrole	578·0	599·2	21·2
Indole	1040	1086·5	46·5
Carbazole	1500	1573·8	73·8
Pyridine	674·7	697·7	23·0
Quinoline	1136·9	1184·2	47·3

groups, and in related compounds these values are followed very closely. In non-conjugated dienes and polyenes the heat of hydrogenation is found to be a simple multiple of that for the analogous compound containing only one unsaturated linkage. With conjugated dienes and polyenes, however, the observed heats of hydrogenation have been found to be less than the calculated values, and less than that of their non-conjugated isomers. The resonance energy of the conjugated compounds is lost in the hydrogenation; and the difference between the observed and calculated values is therefore a measure of the empirical resonance energy (see table 2.3).

TABLE 2.3. *Heats of hydrogenation and empirical resonance energies*

(Wheland, 1955)

Substance	Heat of hydrogenation (kcal mole^{-1}) Calculated	Observed	Empirical resonance energy (kcal mole^{-1})
1,4-Pentadiene	60·6	60·8	—
1,5-Hexadiene	60·6	60·5	—
Limonene	54·9	54·1	—
1,3-Butadiene	60·6	57·1	3·5
1,2-Cyclopentadiene	53·8	50·9	2·9
1,3-Pentadiene	58·3	54·1	4·2
Benzene	85·8	49·8	36·0
Styrene	114·4	77·5	36·9
Indene	109·3	69·9	39·4

It must be emphasized, however, that these empirical resonance energies are not identical with the vertical resonance energies. For benzene, the empirical resonance energy corresponds to the difference between the energy of cyclohexatriene (with alternating bond lengths corresponding to single and double bonds) and benzene (in which all the carbon–carbon bonds are of equal length). The vertical resonance energy is the difference in energy between the structure having alternating double and single bonds of equal length, and benzene (in which the electrons are delocalized). A compression energy is thus involved, and this has been estimated to be 27 kcal mole^{-1}.

In order to calculate the delocalization energy of a molecule or ion it is necessary to calculate the difference between the energy which the π-electrons would have if paired in the simplest possible way to form pure double bonds, and the total energy of the π-electrons in their completely delocalized molecular orbitals.

Consider the allyl radical, $CH_2{=}CH{-}CH_2^{\cdot}$. If it is supposed that two π-electrons form the π-bond between carbon atoms 1 and 2, and that the third electron remains isolated on carbon atom 3, then the total energy of the π-electrons would be $E = 3E_0 + 2\beta$, where E_0 is the energy of an isolated $2p_z$ orbital, and β is the resonance or exchange integral. This constant, β, is indicative of the difference in binding energy between a single and a double bond, between ethane and ethylene: this difference is 2β.

However, the two carbon–carbon bonds in the allyl radical are equivalent, and the electrons occupy non-localized molecular orbitals. Three molecular orbitals are possible, and their energies may be calculated to be $E_0 + \sqrt{2}\beta$, E_0 and $E_0 - \sqrt{2}\beta$. The latter is an anti-bonding orbital for the energy is higher than that of the atomic orbitals. In the ground state two electrons, with anti-parallel spins, occupy the orbital of lowest energy, $E_0 + \sqrt{2}\beta$; and one electron occupies that with energy E_0. The total π-electron energy of the system is therefore $3E_0 + 2\sqrt{2}\beta$. The difference between the energy of the allyl radical assuming it has the classical structure, and its energy assuming the electrons occupy non-localized molecular orbitals, is therefore

$$(3E_0 + 2\beta) - (3E_0 + 2\sqrt{2}\beta), \quad \text{or} \quad -0{\cdot}828\beta;$$

this represents the delocalization energy. As β has a negative value the delocalization energy is positive.

The energy levels for the molecular orbitals in cyclic systems can also be calculated, or derived graphically (Streitwieser, 1961). For a five-membered ring the energy levels are:

$$E_1 = E_0 + 2\beta,$$

$$E_2 = E_0 + 0{\cdot}618\beta,$$

$$E_3 = E_0 + 0{\cdot}618\beta,$$

$$E_4 = E_0 - 1{\cdot}618\beta,$$

$$E_5 = E_0 - 1{\cdot}618\beta.$$

For a six-membered ring the energy levels are:

$$\begin{aligned} E_1 &= E_0+2\beta, \\ E_2 &= E_0+\beta, \\ E_3 &= E_0+\beta, \\ E_4 &= E_0-\beta, \\ E_5 &= E_0-\beta, \\ E_6 &= E_0-2\beta. \end{aligned}$$

And for a seven-membered ring the energy levels are:

$$\begin{aligned} E_1 &= E_0+2\beta, \\ E_2 &= E_0+1{\cdot}247\beta, \\ E_3 &= E_0+1{\cdot}247\beta, \\ E_4 &= E_0-0{\cdot}445\beta, \\ E_5 &= E_0-0{\cdot}445\beta, \\ E_6 &= E_0-1{\cdot}802\beta, \\ E_7 &= E_0-1{\cdot}802\beta. \end{aligned}$$

The delocalization energy can be calculated in terms of β by summing the occupied energy levels times the number of electrons in each energy level, and subtracting 2β for each ethylene unit in one of the Kekulé-type structures.

For the cyclopentadienide anion, therefore, we have:

2 electrons, $2{\cdot}00\beta$ =	$4{\cdot}000\beta$
4 electrons, $0{\cdot}618\beta$ =	$2{\cdot}472\beta$
	$6{\cdot}472\beta$
Minus 2 times 2β	$4{\cdot}000\beta$
D.E.	$2{\cdot}472\beta$

For benzene, we have:

2 electrons, $2{\cdot}00\beta$ =	$4{\cdot}000\beta$
4 electrons, $1{\cdot}00\beta$ =	$4{\cdot}000\beta$
	$8{\cdot}000\beta$
Minus 3 times 2β	$6{\cdot}000\beta$
D.E.	$2{\cdot}000\beta$

For the cycloheptatrienium cation, we have:

$$\begin{array}{lr} 2 \text{ electrons, } 2{\cdot}00\beta = & 4{\cdot}000\beta \\ 4 \text{ electrons, } 1{\cdot}247\beta = & 4{\cdot}988\beta \\ \hline & 8{\cdot}988\beta \\ \text{Minus 3 times } 2\beta & 6{\cdot}000\beta \\ \hline \text{D.E.} & 2{\cdot}988\beta \end{array}$$

And similarly, for the cycloheptatrienide anion, we have:

$$\begin{array}{llr} 2 \text{ electrons,} & 2{\cdot}000\beta = & 4{\cdot}000\beta \\ 4 \text{ electrons,} & 1{\cdot}247\beta = & 4{\cdot}988\beta \\ 2 \text{ electrons,} & -0{\cdot}445\beta = & -0{\cdot}890\beta \\ \hline & & 8{\cdot}098\beta \\ \text{Minus 3 times } 2\beta & & 6{\cdot}000\beta \\ \hline & \text{D.E.} & 2{\cdot}098\beta \end{array}$$

Delocalization energies for some polycyclic benzenoid hydrocarbons have been collected in table 2.4, which includes resonance energy values determined from heats of combustion data, and transferred from table 2.3.

For planar and near-planar benzenoid hydrocarbons, such as those given in table 2.4, the correlation between the delocalization

TABLE 2.4. *Delocalization energies and empirical resonance energies of polycyclic benzenoid hydrocarbons*

(Streitwieser, 1961)

Substance	Molecular orbital delocalization energy (β)	Empirical resonance energy from heat of combustion (kcal mole^{-1})
Benzene	2·000	36·0
Naphthalene	3·683	61·0
Anthracene	5·314	83·5
Phenanthrene	5·448	91·3
Naphthacene	6·932	110·0
Benz[*a*]anthracene	7·101	111·6
Benzo[*c*]phenanthrene	7·187	109·6
Chrysene	7·190	116·5
Triphenylene	7·275	117·7
Pyrene	6·506	108·9
Perylene	8·245	126·3

energies calculated by the molecular orbital method and the empirical resonance energies determined from heats of combustion, is excellent. For non-benzenoid hydrocarbons, and for non-planar benzenoid hydrocarbons, however, the agreement is much less good. With such compounds the empirical resonance energies determined from heats of combustion are lower than the theoretically calculated values. It is not surprising that this should be so. With non-benzenoid hydrocarbons, for example, there is often considerable distortion from the normal sp^2 bond angle of 120°; the theoretically deduced delocalization energy should therefore be reduced by an amount equal to the strain energy.

TABLE 2.5. *Delocalization energies and empirical resonance energies of non-benzenoid hydrocarbons*

Substance	Delocalization energy (β)	Empirical resonance energy (kcal mole^{-1})
Cyclobutadiene	0	—
Cyclopentadienide anion	2·48	—
Fulvene	1·466	—
Fulvalene	2·779	—
Cycloheptatrienium cation	2·99	—
Tropolone	—	21
Cyclo-octatetraene	1·657*	4·8
Pentalene	2·46	—
Azulene	3·364	33
Biphenylene (I)	4·506	17·1

* Calculated for a planar ring. For double bonds completely orthogonal, D.E. = 0.

Empirical resonance energies, and delocalization energies, for some non-benzenoid hydrocarbons (see chapters 3, 4) have been collected in table 2.5, and in every case it can be seen that the value derived from heats of combustion is lower than that expected from the theoretical value. Biphenylene (XCV), for example, has a relatively high delocalization energy of 4·506β, or 0·506β more than two benzene rings; but the empirical resonance energy from the heat of combustion is only 17·1 kcal mole^{-1}.

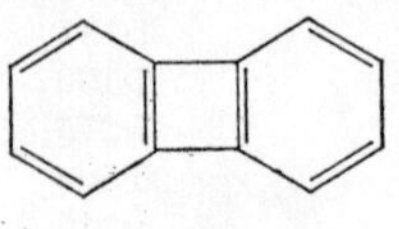

(XCV)

Non-planarity also reduces the resonance energy in benzenoid, in non-benzenoid, and in heterocyclic compounds. Consider, for example, the effect of distortion of a planar benzene molecule by folding the ring through an angle of 2θ around the 1,4 axis (as in XCVI). The effect on the energy levels of the various molecular orbitals can then be estimated (Coulson, 1958). As a first approximation it is reasonable to suppose that the $2p_z$ atomic orbitals at atoms 2,3 remain perpendicular to the plane 1,2,3,4, that those at 5,6 remain perpendicular to the plane 1,4,5,6, and that the orbitals at C_1 and C_4 are directed along the plane of symmetry through C_1C_4. Then

$$\beta_{2,3} = \beta_{5,6} = \beta \quad \text{and} \quad \beta_{1,2} = \beta_{3,4} = \beta_{4,5} = \beta_{1,6} = \beta\cos\theta.$$

If θ is very small the energy levels become

$$E_0 \pm (2 - \tfrac{2}{3}\theta^2)\beta, \quad E_0 \pm (1 - \tfrac{2}{3}\theta^2) \quad \text{and} \quad E_0 \pm \beta.$$

The delocalization energy of non-planar benzene is therefore

$$(4\beta - \tfrac{4}{3}\beta\theta^2 + 2\beta - \tfrac{4}{3}\beta\theta^2 + 2\beta) - 6\beta \quad \text{or} \quad 2\beta(1 - \tfrac{4}{3}\theta^2),$$

instead of 2β.

It must be concluded that the delocalization energy of a non-planar molecule is less than that of the corresponding planar structure, depending on the angle of deformation, θ. If $\theta = 0$, the delocalization energy reverts to the 'normal' value.

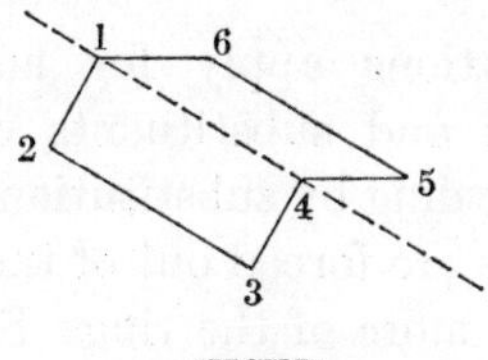

(XCVI)

Many benzenoid compounds are now known which are not absolutely planar (Ferguson & Robertson, 1963). Examples of non-planar structures among polycyclic benzenoid compounds are benzo[*c*]phenanthrene (XCVII), tetrabenzo[*a*,*c*,*f*,*h*]naphthalene (XCVIII), dibenzo[*c*,*g*]phenanthrene (XCIX), and phenanthro[3,4-*c*]phenanthrene, or hexahelicene (C). The non-planarity of dibenzo[*c*,*g*]phenanthrene has been confirmed by X-ray evidence, and by the observed mutarotation of the morphine salt of its 9,10-dicarboxylic acid. Hexahelicene has been resolved, and the *laevo*-isomer found to have a high rotation, $[\alpha]_D - 3640°$. All these compounds are overcrowded in the sense that there is insufficient room for the hydrogen atoms in juxtaposition; and in hexahelicene because even the carbon atoms would overlap if the molecule were not distorted from the planar condition. From the above discussion, the resonance energy of each non-planar hydrocarbon must be less than that predicted for a planar molecule of the same structure.

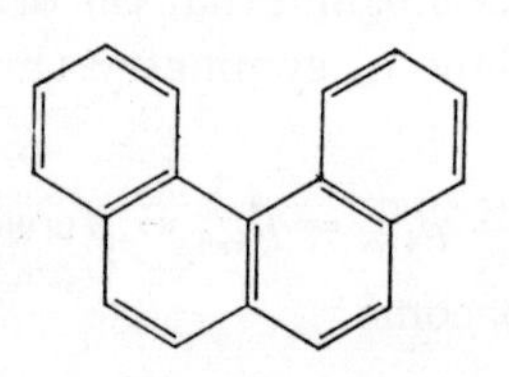

(XCVII)

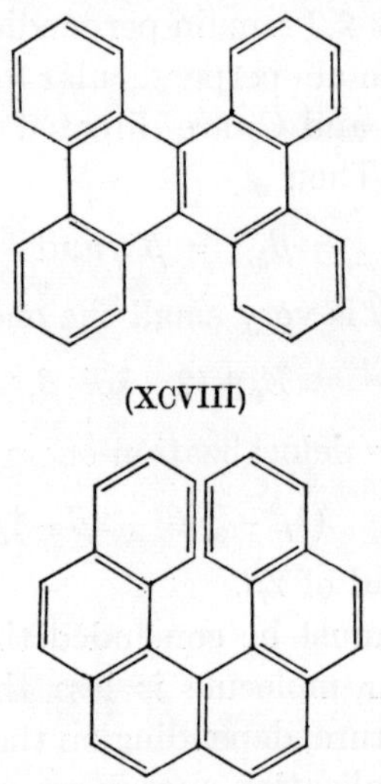

(XCVIII)

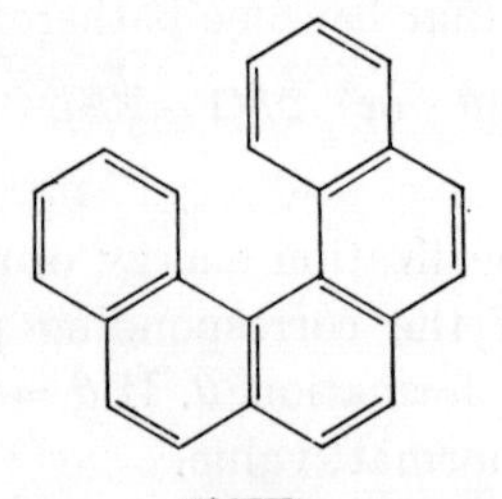

(XCIX)

(C)

The same considerations apply for heterocyclic and non-benzenoid compounds; and substituents can also cause over-crowding. This overcrowding by substitution can be relieved if the substituents themselves are forced out of the plane of the ring, or by distortion of one or more of the rings. Somewhat surprisingly it has been found that relief is often produced by distortion of the ring. In 7,12-dimethylbenz[*a*]anthracene (CI), for example, the 'benz' ring is bent down out of the general plane by about 20°, and the 12-methyl group bent up about 12° (Sayre & Friedlander, 1960). Overcrowding of this nature has a significant effect on the resonance energy. Measurements of heats of combustion have established the strain energies due to methyl interferences as $12{\cdot}6 \pm 1{\cdot}5$ kcal mole^{-1} for 4,5-dimethylphenanthrene and $11{\cdot}0 \pm 1{\cdot}6$ kcal mole^{-1} for 1,12-dimethylbenzo[*c*]phenanthrene (Frisch *et al.* 1963).

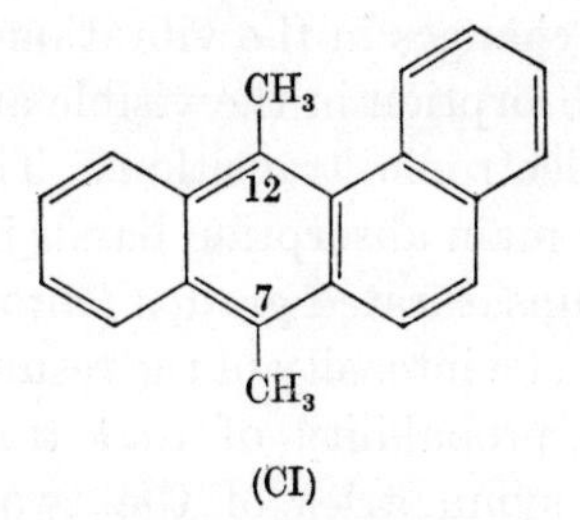

(CI)

It seems therefore that small distortions from planarity can be accommodated without destruction of aromaticity; but few detailed data are available. Molecular overcrowding in non-benzenoid hydrocarbons and related compounds can certainly prevent aromaticity. [10]Annulene and [14]annulene would, if planar, be expected to be aromatic on the basis of Hückel's rule; but overcrowding of the internal hydrogen atoms prevents planarity of the carbon skeletons. [10]Annulene is not a stable compound, and the properties of [14]annulene at room temperature suggest that this compound is also non-aromatic (see chapter 3). Similarly, [18]annulene trioxide is planar and aromatic, as is [18]annulene dioxidesulphide; but [18]annulene oxidedisulphide and [18]annulene trisulphide are non-planar and non-aromatic. In this latter series there is a complete gradation from planarity and aromaticity to non-planarity and non-aromaticity (see chapter 3).

It must be concluded therefore that planarity is not an essential requirement for aromaticity; but the molecule or ion must be sufficiently near planar to allow the π-electrons to occupy bonding molecular orbitals in a closed electron shell.

2.3. Electronic absorption spectra. The absorption and emission spectra of organic compounds are associated with energy changes in the molecules. Absorption of light causes a change from a lower to a higher energy level, and emission of light results from a change from a higher to a lower energy state. The frequency of the light emitted or absorbed is given by $\Delta E = h\nu$, where ΔE is the energy difference between the two levels, h is Planck's constant, and ν is the frequency.

The energy of a molecule may be broadly divided into kinetic (rotational and vibration-rotational) energy, and electronic energy. Changes in rotational energy give absorption bands in the far infrared region, and changes in the vibration-rotational energy in the near infrared. Absorption in the visible and ultraviolet region corresponds with electronic transitions. The simplest organic molecules have their main absorption bands in the far ultraviolet, and molecules with unsaturated groups (chromophores) absorb in the near ultraviolet. The intensity of the resulting absorption band is governed by the probability of that transition, and this is determined by the symmetries of the two states. If the two symmetries are the same, or almost identical, then little or no change in dipole moment accompanies the transition which is therefore forbidden. If the change in dipole moment is large the transition is allowed and the absorption band has a high intensity. Thus $s \rightarrow p$ transitions are allowed, or permitted; but $s \rightarrow s$ transitions are forbidden. However, if s orbitals have their perfect symmetry destroyed by the near presence of other atoms the $s \rightarrow s$ transitions become partially permitted and result in weak absorption.

Ethylene is the simplest unsaturated hydrocarbon and gives rise to a strong absorption band in the 175 nm region, and all compounds having an isolated double bond absorb in this region. This absorption can be readily explained in terms of molecular orbital theory. In ethylene in the ground state, two π-electrons occupy a (bonding) molecular orbital which may be designated π_u, the subscript indicating that the orbital has opposite phases on each side of the node. The (antibonding) molecular orbital of higher energy is unoccupied in the ground state, but an electron can be promoted to this level by absorption of energy. This orbital, designated π_g, has a new nodal plane across the C—C link (see fig. 2.1). The energy difference between the two orbitals (162 kcal mole^{-1}) is such that the absorption band is at 175 nm. Transitions

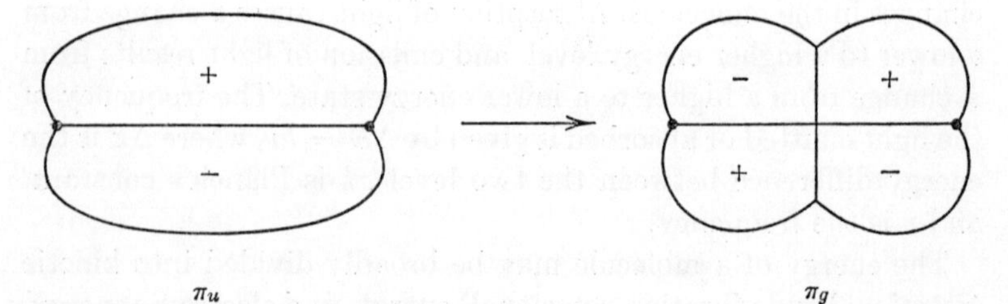

Fig. 2.1. π-Orbitals of ethylene.

of this type are called $\pi \rightarrow \pi^*$, $N \rightarrow V$, or ${}^1B_{1u} \leftarrow {}^1A_g$ transitions. Incidentally, by excitation in this way the energy required to cause rotation about the C=C bond is reduced, and hence *cis–trans* isomerizations are facilitated by irradiation.

The absorption spectra of butadiene and of other conjugated polyenes can be interpreted in a similar fashion. With butadiene, the π-electrons can occupy four molecular orbitals. In the ground state two electrons occupy the molecular orbital (Ψ_1) of lowest energy, and two occupy the molecular orbital (Ψ_2) of slightly higher energy. Electrons are promoted to molecular orbitals (Ψ_3, Ψ_4) of higher energy by absorption of light. The different orbitals have different symmetries (fig. 2.2), and as the energy difference between Ψ_2 and Ψ_3 is less than that for the excitation of an electron from one molecular orbital to another in ethylene, butadiene absorbs at longer wavelengths than ethylene. Experimentally, it is found that butadiene absorbs with a maximum at 217 nm, and polyenes absorb at even longer wavelengths. Compounds having four or more conjugated double bonds have absorption which extends into the visible region, and these compounds are therefore coloured (see table 2.6).

TABLE 2.6. *Electronic absorption bands of polyene systems*

Compound	Number of double bonds	λ_{max} (nm)	ϵ_{max}	Colour
Ethylene	1	175	*ca.* 5000	Colourless
Butadiene	2	217	21000	Colourless
Hexatriene	3	258	35000	Colourless
Dimethyloctatetraene	4	296	52000	Pale yellow
Decapentaene	5	335	118000	Pale yellow
Dimethyldodecahexaene	6	360	70000	Yellow
Dihydro-β-carotene	8	415	210000	Orange
Lycopene	11	470	185000	Red
Dehydrolycopene	15	504	150000	Violet

The absorption spectra of cyclic polyenes are similar to those of the open-chain analogues, but the intensity of absorption is much less. For example, although butadiene shows absorption at 217 nm with $\epsilon = 21{,}000$, cyclopentadiene absorbs at 238·5 nm with $\epsilon = 3400$; and 1,3-cyclohexadiene absorbs at 256 nm with $\epsilon = 7950$.

It will be useful at this stage to compare 1,3,5-hexatriene and benzene. Hexatriene has six π-electrons, so that six molecular orbitals are possible. These have energies as follows:

$$
\begin{aligned}
E_1 &= E_0 + 1{\cdot}802\beta,\\
E_2 &= E_0 + 1{\cdot}247\beta,\\
E_3 &= E_0 + 0{\cdot}445\beta,\\
E_4 &= E_0 - 0{\cdot}445\beta,\\
E_5 &= E_0 - 1{\cdot}247\beta,\\
E_6 &= E_0 - 1{\cdot}802\beta.
\end{aligned}
$$

In the ground state two electrons occupy molecular orbital Ψ_1, energy E_1; two occupy orbital Ψ_2, energy E_2; and two occupy orbital Ψ_3, energy E_3. The first excited (singlet) state is obtained by promotion of an electron from the orbital of highest energy to the lowest unoccupied orbital. Experimentally, it is found that in iso-octane solution, hexatriene absorbs at 268 nm, with an extinction coefficient ϵ of 34,600 (Jaffé & Orchin, 1962).

Benzene also has six π-electrons, and six molecular orbitals are possible. These orbitals have energies as follows:

$$
\begin{aligned}
E_1 &= E_0 + 2\beta,\\
E_2 &= E_0 + \beta,\\
E_3 &= E_0 + \beta,\\
E_4 &= E_0 - \beta,\\
E_5 &= E_0 - \beta,\\
E_6 &= E_0 - 2\beta.
\end{aligned}
$$

In the ground state two electrons occupy molecular orbital Ψ_1 with energy E_1; two occupy orbital Ψ_2, energy E_2; and two occupy orbital Ψ_3, energy E_3. Orbitals Ψ_2 and Ψ_3 are degenerate. Each wave function is distinct: both have the same energy and the same number of nodal planes; but the orientation of the nodal planes differs. Experimentally, it is found that benzene gives three regions of absorption in the ultraviolet (see table 2.7).

The Group III bands, which occur in the 230–260 nm region, correspond to a forbidden transition; absorption in this region has low intensity (ϵ about 200). The effect of vibration can be clearly seen in the well-defined sub-peaks, the spacing of which corresponds

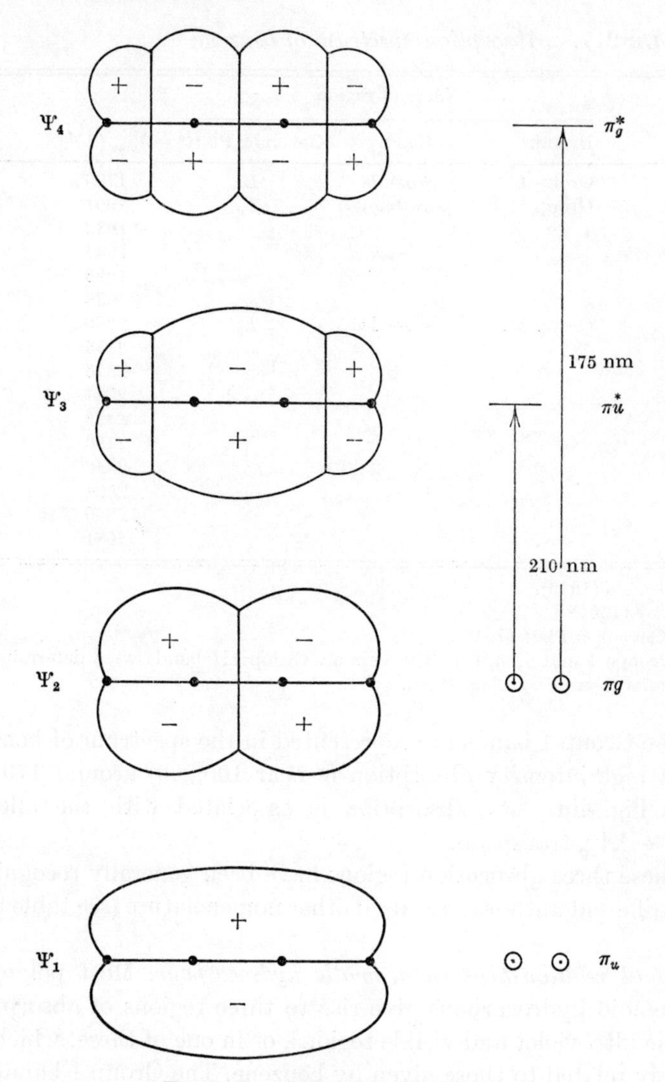

Fig. 2.2. π-Orbitals in butadiene

to the vibrational frequency. The Group III absorption bands are believed to be due to a ${}^1B_{2u} \leftarrow {}^1A_{1g}$ transition.

The Group II bands are medium-intensity bands (ε near 10,000) near 200 nm. These bands are associated with a transition from the highest occupied to the lowest unoccupied molecular orbital; that is, to the forbidden ${}^1B_{1u} \leftarrow {}^1A_{1g}$ transition.

TABLE 2.7. *Absorption spectrum of benzene*

Name of region			
Braude[a]	Clar[b]	Klevens & Platt[c]	λ_{max} (Å)[d]
Group I	β-bands	B_b	1790
Group II	*para*-bands	L_a	1901
			1932
			1964
			1998
			2038
Group III	α-bands	L_b	2290
			2335
			2376
			2387
			2428
			2486
			2542
			2604
			2640
			2681

[a] Braude (1945).
[b] Clar (1964).
[c] Klevens & Platt (1949).
[d] Groups I and II are for the vapour. Group III bands were determined in methanol-ethanol by Clar (1964).

The Group I bands are represented in the spectrum of benzene by a high-intensity absorption (ϵ near 100,000) around 179 nm. This high-intensity absorption is associated with the allowed ${}^1E_{2u} \leftarrow {}^1A_{1g}$ transition.

These three absorption regions have been generally recognized, but different authors have used other nomenclature (see table 2.7).

Spectral relationships in aromatic hydrocarbons. Most polycyclic benzenoid hydrocarbons give rise to three regions of absorption, in the ultraviolet and visible regions, or in one of these, which are closely related to those given by benzene. The Group I bands are high-intensity bands with $\log \epsilon$ 4·5–5·2, and correspond with the absorption band at 1790 Å given by benzene. The Group II bands have $\log \epsilon$ around 3·6–4·1; and the Group III bands are low-intensity bands with $\log \epsilon$ 2·3–3·2.

Extensive studies of the spectral relationships of polycyclic compounds have been published. Clar (1964) has found that the three regions of absorption are closely related to those given by benzene and that they move to longer wavelengths on linear or angular annellation; moreover the shifts to longer wavelengths follow simple rules. In the polyacenes, for example, the Groups I, II and III bands all move to longer wavelengths as the number of rings increases, but the Group II bands show the greatest shifts per benzene ring. The first maximum in each group of absorption bands for benzene is at 1790, 2038 and 2604 Å respectively. For naphthalene, the wavelengths of the corresponding bands are 2210, 2850 and 3110 Å. With anthracene, however, the Group I bands occur at 2515 Å (first maximum) and the first maximum in the Group II bands occurs at 3745 Å; but the low-intensity Group III bands are now hidden under the medium-intensity Group II bands. This absorption maximum for anthracene is just outside the visible region; but substituents which are strongly conjugated with the ring system (for example, at the 9,10 positions) cause a bathochromic shift of the absorption bands. Such substituted anthracenes absorb into the visible region and are therefore pale yellow or yellow. Similarly the first maximum in the Group II absorption region of naphthacene occurs at 4710 Å, and this compound is orange. The absorption bands of the higher benzologues, pentacene and hexacene, are displaced to even longer wavelengths, and these compounds are violet-blue and black-green respectively (see table 2.8).

It is usually possible to discern three main regions of absorption in the spectra of angular polycyclics as well as in the polyacenes.

TABLE 2.8. *Absorption spectra of polyacenes*

(Clar, 1964)

Compound	Position of first maxima (Å)			Colour of compound
	Group I	Group II	Group III	
Benzene	1790[a]	2038[a]	2604[b]	Colourless
Naphthalene	2210[b]	2850[b]	3110[b]	Colourless
Anthracene	2515[b]	3745[b]	—	Colourless
Naphthacene	2740[c]	4710[c]	—	Orange
Pentacene	3030[d]	5820[e]	—	Deep violet-blue
Hexacene	—	—	—	Deep black-green
Heptacene	—	—	—	Deep green-black

[a] Vapour.
[b] In methanol-ethanol.
[c] In ethanol.
[d] In benzene.
[e] In trichlorobenzene.

Here again the bands shift to longer wavelengths with increasing numbers of benzene rings; but with the angular polycyclics the Group II bands show approximately the same shift per benzene ring as the Group I and Group III bands. While the first maximum in each group of absorption bands for benzene occurs at 1790, 2038 and 2604 Å, those for the corresponding bands for phenanthrene are at 2510, 2925 and 3450 Å, and those for pentaphene are at 3145, 3560 and 4235 Å.

Absorption spectra of aromatic heterocyclic compounds. The electronic structure of pyridine resembles that of benzene very closely, and it might be expected that the electronic absorption spectrum of pyridine would be similar to that of benzene. This is the case: pyridine has three regions of absorption, around 170, 200 and 250 nm. Similarly quinoline and isoquinoline show the same three main regions of absorption as naphthalene.

Electronically, pyridine differs from benzene in having two non-bonding or n-electrons which can interact with polar solvent molecules. In polar solvents pyridine shows no discernible absorption corresponding to $n \rightarrow \pi^*$ transitions; but in non-polar solvents such transitions produce a long tail to the $\pi \rightarrow \pi^*$ absorption, and the spectrum of the vapour also shows the expected absorption partially submerged under the $\pi \rightarrow \pi^*$ absorption.

Diazines also show three main regions of absorption, especially in ethanol solution. In addition, however, most diazines (especially in non-polar solvents) give an absorption band at longer wavelengths which is associated with a transition of one of the lone-pair electrons to a π^* orbital. The spectra of pyridazine, pyrimidine and pyrazine in cyclohexane, for example, all show absorption around 250 nm corresponding to the Group III bands of benzene; but in addition they absorb around 290–360 nm, due to $n \rightarrow \pi^*$ transitions. Similarly, cinnoline in cyclohexane has absorption bands below 230 nm which clearly correspond to the Group I bands given by naphthalene, two maxima at 275·5 and 286 nm which correspond to the Group II bands, and three maxima at 308·5, 317 and 322·5 nm which correspond to the Group III region. In addition, cinnoline gives rise to low-intensity absorption, with a maximum at 390 nm, due to $n \rightarrow \pi^*$ transitions.

The absorption spectra of thiophen, pyrrole and of furan resemble the spectrum of benzene less closely. Thiophen does give three regions of absorption, at 140–160 nm, at about 180–200 nm and at 210–240 nm. Pyrrole has an absorption band at about 210 nm with absorption extending, with no fine structure, to about 260 nm, and further bands at 183 and 172 nm. The vapour spectrum of furan shows a virtually structureless band at about 205 nm and extending only to about 230 nm. There is also a band with a maximum about 191 nm having considerable vibrational structure.

With benzo-derivatives of five-membered heterocyclic compounds the situation is much clearer. The spectra of benzofuran, indole and thianaphthene have been compared. They have some features in common, and all three spectra show some resemblance to that given by naphthalene. Only thianaphthene shows three well-defined regions of absorption: around 227, 245–270 and 280–300 nm. Only two regions could be resolved with indole at 219 nm (Group I) and 288 nm (Group III). With benzofuran the maximum at 244 nm probably corresponds with the Group II region in the naphthalene spectrum, and at 281 nm with the Group III region.

The spectra of the tricyclic compounds, dibenzothiophen, carbazole and dibenzofuran resemble that of phenanthrene. The main difference is that the Group I bands given by the heterocyclic compounds are shifted to shorter wavelengths, and the Group III bands are more intense.

2.4. Induced ring currents. Almost all stable organic molecules exhibit the phenomenon of diamagnetism; in diamagnetic substances a magnetic field induces small circuit currents whose magnetic fields oppose the inducing field. Diamagnetism occurs in isolated atoms, the electrons being given a circular motion in planes perpendicular to the field, but the effect is larger in polyatomic molecules. Each electron makes its own contribution to the induced magnetic polarization, the contribution being proportional to the mean area of the orbit which it describes. In saturated molecules, or in molecules which contain only isolated centres of unsaturation, the electrons are relatively isolated and the magnetic field causes them to circulate only within the atoms or bonds to which they belong (Selwood, 1956; Ingold, 1953).

The magnetic susceptibility per c.c. can be determined experimentally; and this, divided by the density and multiplied by the molecular weight, gives the molecular susceptibility, χ. Diamagnetic spherical atoms, and a few special classes of molecule, are magnetically isotropic, their magnetic susceptibilities being the same in all directions. Most diamagnetic molecules, however, are not isotropic, their measured magnetic susceptibilities being average values related to the three principal magnetic susceptibilities, directed along three mutually perpendicular, principal axes of magnetic susceptibility.

The molecular magnetic susceptibility of a diamagnetic substance is negative; it is therefore convenient to refer to $-\chi$, the molecular diamagnetic susceptibility. Furthermore, as diamagnetic susceptibilities are of the order of 10^{-6} c.g.s. electromagnetic units per mole, the practical unit is often taken as $-10^6\chi$.

About 1910 Pascal found that the molecular susceptibility of a diamagnetic susbtance is an additive property of its component atoms and bonds. Tables of diamagnetic constants for atoms and bonds have been compiled and, using these values, the molecular susceptibility of a saturated molecule, or a molecule having only isolated centres of unsaturation, can be calculated with surprising accuracy. Some of Pascal's constants, as modified by Ingold, are given in table 2.9.

TABLE 2.9. *Diamagnetic constants for atoms and multiple bonds* (Ingold, 1953)

Atom	$-10^6\chi$	Bond corrections	$-10^6\chi$
H	2·9	C=C	−5·5
C	6·0	C≡N	−8·2
N	5·6	C=O	−6·3
O	4·6		
S	15·0		

With conjugated molecules, however, the observed molecular diamagnetic susceptibilities have been found to differ from the values calculated by summation of the requisite Pascal's constants. This difference, called the exaltation ($-\Delta\chi$), is small with non-aromatic compounds, but is relatively large for aromatic compounds. The susceptibility exaltation (in $-10^6\chi$) for butadiene is +0·5, but for benzene is +18·0 and for pyridine is +18·5. High

TABLE 2.10. *Susceptibility exaltations in hydrocarbons* (Craig, 1959*a*)

Molecule	$-10^6\chi$ (Experimental)	Pascal sum	$-\Delta\chi$	$-\Delta\chi/n$
Benzene	55·6	36·9	18·7	18·7
Naphthalene	91·9	55·7	36·2	18·1
Anthracene	129·4	74·5	54·9	18·3
Phenanthrene	127·9	74·5	53·4	17·8
Chrysene	160·7	93·3	67·4	16·9
Biphenyl	102·9	68·0	34·9	17·5
Tropone	54	41·2	12·8	
Tropolone	61	45·8	15·2	
Cyclo-octatetraene	51·9	49·2	2·7	
Azulene	91	55·7	35·3	

values are also found for furan (+14) and for thiophen (+18). Other values have been collected in table 2.10 which also gives, where applicable, the values for $-\Delta\chi/n$, where n is the number of benzenoid rings in a polycyclic benzenoid hydrocarbon. In these condensed systems it is seen that the exaltation increases almost linearly with the number of benzenoid rings. Non-benzenoid aromatic compounds also exhibit marked exaltations (azulene, +35·3; tropolone, +15·2), but cyclic polyenes do not (cyclo-octatetraene, +2·7). The susceptibility exaltation for benzene is +3·1 per carbon atom; that for naphthalene is +3·6 and for azulene 3·5 per carbon atom (Craig, 1959*a*). The marked difference between benzene and cyclo-octatetraene was noted in 1948 (Pink & Ubbelohde) and provided the first clear physical evidence of the lack of aromaticity in the latter hydrocarbon.

The susceptibility exaltations found for aromatic compounds are diagnostic, and can be explained in terms of the electronic structure associated with aromaticity. In aromatic molecules the π-electrons are not confined to individual atoms or bonds, but are relatively free. The diamagnetic currents are therefore not limited to atoms, but also involve an extended orbit around the ring. This induced ring current which is, of course, much greater than the small circuit currents associated with the σ-electrons, produces a secondary magnetic field such that the applied force is opposed within the ring and reinforced outside the ring (fig. 2.3). The induced ring current is responsible for the large diamagnetic anisotropy exhibited by aromatic molecules.

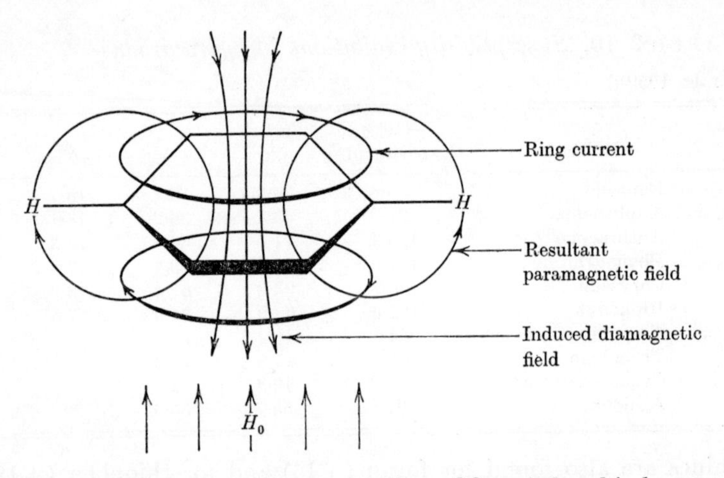

Fig. 2.3. Ring current and magnetic lines of force induced in benzene by an applied magnetic field.

The induced ring current may also be said to be diagnostic, and Elvidge & Jackman (1961) have defined an aromatic compound *as a compound which will sustain an induced ring current.*

This property of aromatic compounds has become increasingly important in recent years as induced ring currents can often be detected experimentally in a simple fashion by determination of the proton magnetic resonance spectrum (Jackman, 1959).

In the p.m.r. spectrum the vinylic protons of non-conjugated olefins usually absorb in the range $\tau 4{\cdot}3$–$5{\cdot}4$; and the protons of cyclic conjugated polyolefins absorb around $\tau 4{\cdot}1$–$4{\cdot}6$. The protons of the non-aromatic cyclo-octatetraene, for example, absorb at $\tau 4{\cdot}31$. With benzene, however, the protons (which are, of course, outside the ring) are strongly deshielded by the secondary magnetic field, and absorb at $\tau 2{\cdot}81$. On the other hand, a proton which is held somehow in the neighbourhood of the six-fold symmetry axis is shielded by the ring current, and such protons absorb at higher field than normal. With hydrogen-bonded complexes of benzene and chloroform, for example, the chloroform proton is located in a shielded position (CII), and the chloroform resonance in benzene is shifted about 1·56 p.p.m. to high field compared with chloroform in cyclohexane. Similarly, in the polymethylenebenzene (CIII), the two central methylene groups are abnormally shielded and these protons absorb at higher field (Waugh & Fessenden, 1957).

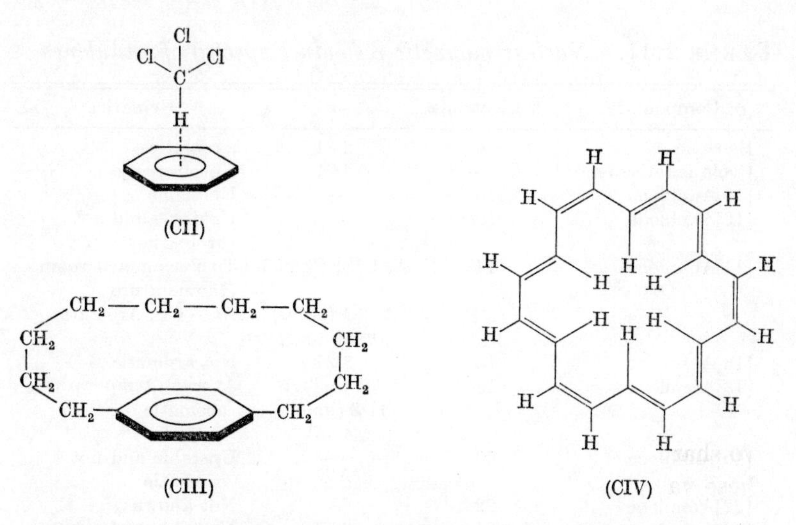

Some macrocyclic aromatic hydrocarbons (see chapter 3) possess not only protons which are external to the ring, but protons which are inside the ring. With [18]annulene (CIV), for example, there are twelve external protons which (like those in benzene) are strongly deshielded. In addition, there are six internal protons which are strongly shielded. These shielding and deshielding effects are reflected in the chemical shifts: the external protons absorb at unusually low fields, and the internal protons at unusually high fields. The external protons of [18]annulene absorb at $\tau 1{\cdot}1$ and the internal protons at $\tau 11{\cdot}8$, the area under the first peak being twice that under the second. A ring current in [18]annulene can be inferred with confidence, and the aromaticity of this macrocyclic compound is thereby established.

Other annulenes have also been studied by the n.m.r. spectroscopic method, and the results have been summarized in table 2.11; more details are given in chapter 3. With some annulenes it is found that the external protons give rise to a peak or peaks at very low field, and the internal protons to a peak or peaks at very high field. In these molecules a ring current can be inferred, and the aromaticity established. On the other hand, some annulenes give a peak or peaks in the olefin region, and no difference between internal and external protons can be observed. The absence of a ring current can therefore be inferred, and such compounds must be regarded as cyclic polyolefins.

TABLE 2.11. *Nuclear magnetic resonance spectra of annulenes*

Compound	π-Electrons	τ	Remarks
Benzene	6	2·81	Aromatic
Cyclo-octatetraene	8	4·31	Not aromatic
[10]Annulene	10	—	Unstable
[12]Annulene	12	—	Unstable and not aromatic
[14]Annulene	14	A, 4·42; B, 3·93	Two isomers at room temperature
		A, 2·4 (outer) 10·0 (inner)	At −60°. Aromatic
[16]Annulene	16	3·27	Not aromatic
[18]Annulene	18	1·1 (outer) 11·8 (inner)	At room temperature; aromatic
		4·55	At 110°
[20]Annulene	20	—	Unstable and not aromatic
[22]Annulene	22	—	Not known
[24]Annulene	24	3·16	Unstable and not aromatic
[26]Annulene	26	—	Not known
[28]Annulene	28	—	Not known
[30]Annulene	30	—	Unstable and not aromatic

Even so, the n.m.r. spectrum must be interpreted with caution. [14]Annulene (CV) is not completely planar, as the internal protons are overcrowded, and two conformers of this compound are known. The n.m.r. spectrum of [14]annulene at room temperature shows two sharp singlets at τ4·42 and τ3·93, due to the two conformers. These values would seem to indicate a polyolefin having no ring current and no aromaticity. However, the changes which occur when the solution is cooled indicate that this conclusion may not be valid. When the solution is cooled the τ4·4 band (isomer A) has been found to broaden, and with further cooling this band disappears, being replaced by a very broad new band at τ2·7. At −60° the spectrum consists of broad peaks at τ2·4 and 10·0 due to isomer A, and at τ3·9 due to isomer B. The τ2·4 peak has been assigned to the ten outer protons, and the τ10·0 band to the four inner protons. At low temperatures a ring current may therefore be inferred; and it is possible that [14]annulene possesses a ring current even at room temperature—the singlets at τ4·42 and 3·93 may be averages for the chemical shifts of the inner and outer protons, due to rapid interchange of proton positions.

A somewhat similar situation has been observed with [18]annulene (CIV). The room-temperature spectrum shows two bands, at τ1·1 and τ11·8. At low temperatures the peaks become sharper and exhibit fine structure, and the separation between the bands is increased. At higher temperatures (110°), however, the spectrum consists of a relatively sharp singlet at τ4·55.

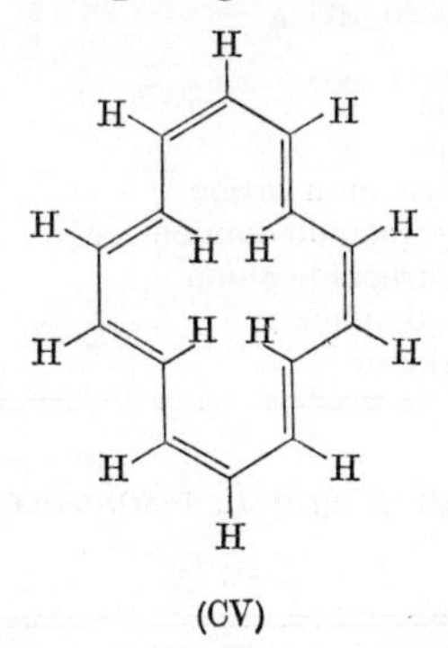

(CV)

The observed chemical shift for an aromatic annulene must clearly depend on the size of the ring, that is, on the distances of the protons from the centre of the ring. (And it may be noted that the susceptibility exaltation for an aromatic compound must also be dependent on the ring-size.)

Other factors also affect the chemical shift. In aromatic ions, the higher the electron density the greater the shielding; and shifts due to changes in ring size are small compared with the effect of electronic charge. The protons in the cyclopentadienide anion (CVI) absorb at τ4·66; the protons of benzene absorb at τ2·81; and the protons of the cycloheptatrienium cation (CVII) at 0·91 (see table 2.12). In some aromatic ions (cyclopropenium cation) the effect due to the ring current is smaller than that due to the ring size or to the charge (Wiberg & Nist, 1961).

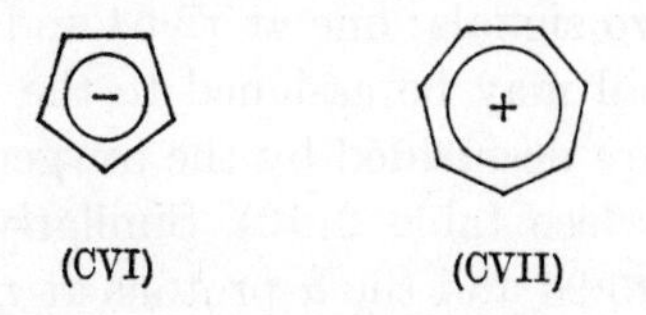

(CVI) (CVII)

With hetero-aromatic compounds the chemical shift is affected by the hetero-atom in addition to the effect of changes in ring-size compared with benzene. With furan, for example, the n.m.r.

TABLE 2.12. *Nuclear magnetic resonance spectra of aromatic ions and related substances*

Molecule or ion	τ
Dipropylcyclopropenium cation	−0·27
Cyclopentadienide anion	4·66
Diazocyclopentadiene	2·8–4·0
Ferrocene	6·01
Ruthenocene	5·61
Magnesocene	4·30
Cycloheptatrienium cation	0·91
Cyclo-octatetraenide dianion	4·38
Cyclononatetraenide anion	3·18
Pentalenide dianion	4·27, 5·02
3-Phenylsydnone	3·22

TABLE 2.13. *Nuclear magnetic resonance spectra of some hetero-aromatic compounds*

Molecule	Position	τ
Furan	α-	2·64
	β-	3·72
Pyrrole	α-	3·47
	β-	3·94
Thiophen	α-	2·84
	β-	2·94
Pyridine	2-	1·49
	3-	2·93
	4-	2·52
Pyridazine	2-	0·83
	3-	2·52
Pyrimidine	2-	0·75
	4,6-	1·22
	5-	2·63
Porphin	*meso*-	−1·22
	β-	0·08
	NH	14·4

spectrum shows two signals: one at τ2·64 and the other at τ3·72. The high-field signal may be assigned to the β-protons; and the α-protons (which are deshielded by the oxygen atom) give rise to the band at τ2·64 (see table 2.13). Similarly, the β-protons of pyrrole absorb at τ3·94 and the α-protons at τ3·47. Special attention may be directed to the macrocyclic heteroaromatic compound, porphin, and to the porphyrins. These compounds are discussed in chapter 3, but it may be noted at this stage that the *meso* and β-protons of porphin are strongly deshielded and give signals at very low field. The internal protons, the protons attached to nitrogen, are strongly shielded, and these protons give a band at τ14·4, which is one of the highest values observed for hydrogen atoms in organic compounds.

With the six-membered hetero-aromatic systems the electronegative effect of the hetero-atom is again important. With pyridine the 2-proton gives a signal at τ1·49, the 3-proton at τ2·93 and the 4-proton at τ2·52. With pyridazine, the 2-proton absorbs at τ0·83 and the 3-proton at τ2·5, and there is a considerable concentration effect. It is also of interest that a substantial ring current has been demonstrated in the boron–nitrogen analogue (CVIII) of naphthalene.

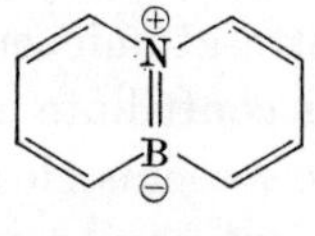

(CVIII)

Chemical shifts of heteroaromatic compounds are often strongly concentration-dependent, and the variation has been shown to depend on the environment within any given molecule. In indoles, for example, the protons in the 2-position vary far more than those in the 3-position.

In an early paper, Elvidge & Jackman (1961) suggested that the magnitude of the ring current might be used as a measure of the aromaticity of a compound; and their studies with 2-pyridones led them to the conclusion that these compounds possess about 35 % of the aromaticity of benzene.

Later, an attempt was made to compare the ring currents in five-membered heterocyclic compounds with that of benzene (Abraham *et al.* 1965). The chemical shifts were compared with those of similarly situated protons in molecules in which no ring current can occur. Taking these observed chemical-shift differences between furan and benzene, for example, and taking into account the areas of the two rings, and the distances of the protons from the centres of the rings, the ratio of the ring currents was deduced. It was concluded that the aromatic ring currents in furan and thiophen do not differ significantly from that in benzene. This conclusion has, however, been disputed (Elvidge, 1965). The result depends on the models chosen for determination of the chemical-shift differences; and it has been claimed that the non-aromatic models used are unsuitable, and indeed that a correct non-aromatic model is strictly not attainable. An alternative approach gives the following relative aromaticities: furan, 0·46; thiophen, 0·75; and pyrrole, 0·59. The attempts to deduce aromaticity from chemical shifts have been further examined in a later paper (Abraham & Thomas, 1966).

Small diamagnetic anisotropies have been observed for some acyclic molecules in which the possibility of ring currents is excluded, and it now seems that the anisotropy of aromatic compounds is a compound effect: part of the observed anisotropy is due to the induced ring current, and part to the local atomic contributions representing intra-atomic currents. It has been estimated that the local terms contribute approximately 30 % to the observed anisotropy. It now seems clear therefore that although the presence of a ring current can be used as a sensitive test for aromaticity, its magnitude cannot be used as a measure of the *degree* of aromaticity.

3. Non-benzenoid hydrocarbons

3.1. Hückel's rule. In the early history of organic chemistry the peculiar stability of benzene and of benzenoid compounds posed a theoretical problem of compelling interest. The significance of the sextet of electrons was first recognized in terms of valencies or residual affinities by Bamberger (1891), and it was more clearly recognized and expressed in terms of electrons by Armit & Robinson (1925). However the cyclopentadienide anion also has six π-electrons, and so has the cycloheptatrienium cation, and it is necessary to enquire whether these ions are aromatic.

The question then arises whether aromatic character or aromaticity is associated only with molecules and ions having a sextet of electrons. Monocyclic structures can be devised which have 2, 4, 6, 8, 10, 12, 16, 18 or more π-electrons; and analogous bicyclic and polycyclic structures can also be conceived. The problem is to decide which of these molecules and ions will be aromatic, and which will be cyclic polyenes exhibiting only limited conjugation. In this respect Hückel's rule and Craig's rule are of great value.

During 1931–1938 Hückel developed the basic pattern of molecular orbital theory of unsaturated and aromatic compounds. He concluded that for any planar monocyclic system the binding energy would vary with the number of π-electrons, and that systems containing $(4n+2)$ π-electrons would be particularly favoured.

According to Hückel's rule, monocyclic coplanar systems of trigonally hybridized atoms which contain $(4n+2)$ π-electrons possess relative electronic stability. This means that planar systems having 2, 6, 10, 14, 18 or 22 π-electrons may be expected to be aromatic. The cyclopropenium cation, the cyclopentadienide anion, the cycloheptatrienium cation, [14]annulene and [18]annulene, for example, should therefore be aromatic provided they are planar; but cyclobutadiene, cyclo-octatetraene, and [12]annulene should be non-aromatic even if they are planar molecules.

Further theoretical work, by Longuet-Higgins & Salem (1959, 1960) led to the prediction that as the rings become very large the $(4n+2)$ rule will no longer operate, but that such ring systems will consist of alternate single and double bonds.

Hückel's rule is of limited applicability as it is concerned only with monocyclic systems. It provides no clue concerning the possible aromatic nature of bicyclic and polycyclic non-benzenoid hydrocarbons which have classical structures with alternate single and double bonds. Craig (1959*b*) has proposed an empirical rule which has proved of value in this connection; but the rule applies only to hydrocarbons in which at least two centres lie on a symmetry axis that converts one Kekulé-type structure to another. The structural formula is first labelled with equal numbers of spin symbols α and β, as far as possible alternately, different symbols being given to the ends of all the double bonds in the Kekulé-type structure. The sum is then taken of the number (f) of symmetrically related π-centres not on the symmetry axis, and the number (g) of interconversions of α and β by rotation about the axis. If this sum, $f+g$, is even, the valence-bond ground state is symmetric and the compound may be expected to be aromatic. If the sum is odd, the compound may be expected to be non-aromatic.

In pentalene (CIX), for example, three pairs of centres are related by the symmetry axis, and $f = 3$. Upon rotation no interconversions of α and β occur; i.e. $g = 0$. Hence $f+g = 3$, an odd number, and pentalene should be non-aromatic. Similarly, for heptalene (CX), $f = 5$ and $g = 0$ so $f+g = 5$, and this hydrocarbon should be non-aromatic; and for azulene (CXI) $f = 4$ and $g = 0$ so that $f+g = 4$, and this hydrocarbon would be expected to be aromatic. Experimentally, heptalene has been found to be an unstable non-aromatic hydrocarbon; and azulene is aromatic.

Theoretical quantities such as bond orders, free valence numbers and π-electron densities for non-benzenoid hydrocarbons can be calculated just as for benzenoid hydrocarbons. One notable difference emerges: in benzene and in unsubstituted polycyclic benzenoid hydrocarbons the π-electron density at each carbon atom is 1·0; in non-benzenoid aromatic hydrocarbons, however, the π-electron densities are sometimes greater, and sometimes smaller, than 1·0. The π-electron densities and bond orders for azulene (CXII) illustrate these points.

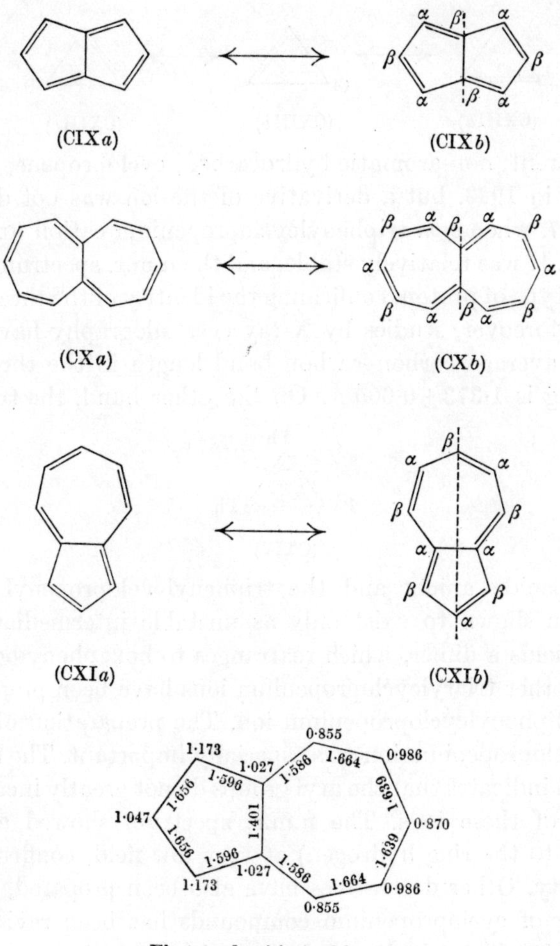

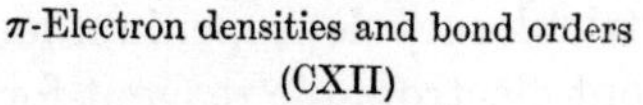

π-Electron densities and bond orders

(CXII)

3.2. 2π-Electron systems.

Cyclopropenium cation. In 1952 it was pointed out (Roberts, Streitwieser & Regan, 1952) that the cyclopropenium cation (CXIII) is, according to Hückel's rule ($4n+2 = 2$; $n = 0$), the simplest possible aromatic system. This cation has a large delocalization energy (2β), but it is impossible to predict *a priori* whether the strain energy would counterbalance, or more than counterbalance, the delocalization energy.

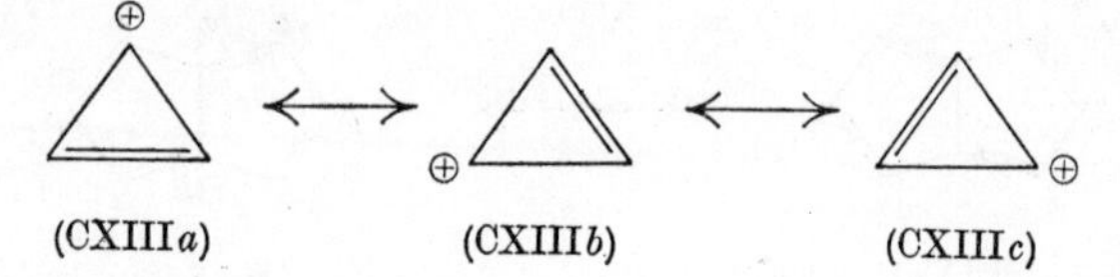

(CXIII*a*) (CXIII*b*) (CXIII*c*)

The parent, non-aromatic hydrocarbon, cyclopropene, was first prepared in 1923, but a derivative of the ion was not described until 1957, when *sym*-triphenylcyclopropenium cation (CXIV) was obtained. It was relatively stable, and the n.m.r. spectrum showed only one type of proton, confirming the identity of the three phenyl groups. Moreover, studies by X-ray crystallography have shown that the average carbon–carbon bond length in the three-membered ring is $1{\cdot}373 \pm 0{\cdot}005$ Å. On the other hand, the triphenyl-

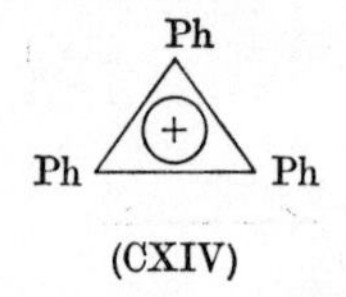

(CXIV)

cyclopropenide anion, and the triphenylcyclopropenyl radical, have been shown to exist only as unstable intermediates. The radical yields a dimer, which rearranges to hexaphenylbenzene.

Many other triarylcyclopropenium ions have been prepared, as has the diphenylcyclopropenium ion. The preparation of the dipropylcyclopropenium ion was especially important. The stability of this ion indicates that the aryl groups do not greatly increase the stability of these ions. The n.m.r. spectrum showed a singlet (assigned to the ring hydrogen) at very low field, confirming the aromaticity. Other derivatives have also been prepared, and the chemistry of cyclopropenium compounds has been reviewed by Krebs (1965). The trichlorocyclopropenium ion (CXV) is of special interest. It is a potent electrophilic reagent for the substitution of aromatic compounds. Reaction with benzene, for example, gives either (CXVI) or (CXVII) depending on the conditions. The latter reacts with water to give diphenylcyclopropenone (CXVIII), and this is a very convenient method for the preparation of this compound.

Cyclopropenone. Cyclopropenone and its derivatives can be represented as cyclopropenium ions, and the preparation of diphenylcyclopropenone (CXIX) was therefore of considerable interest. It

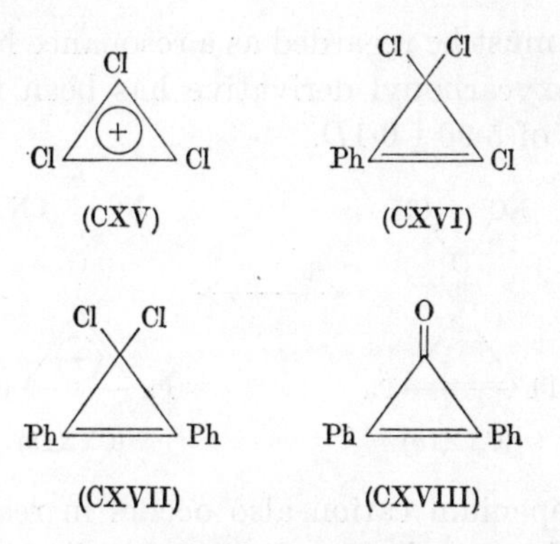

was found to be a remarkably stable compound: it required heating to 130–140° to lose carbon monoxide and form diphenylacetylene. The resonance structure (CXIX*b*) implies that diphenylcyclopropenone would be expected to have a large dipole moment; experimentally, diphenylcyclopropenone was found to have a dipole moment of 5·08*D*, which is much larger than those of other ketones (benzophenone, 2·97*D*; acetone, 2·8*D*). It is also noteworthy that diphenylcyclopropenone behaves as a base and

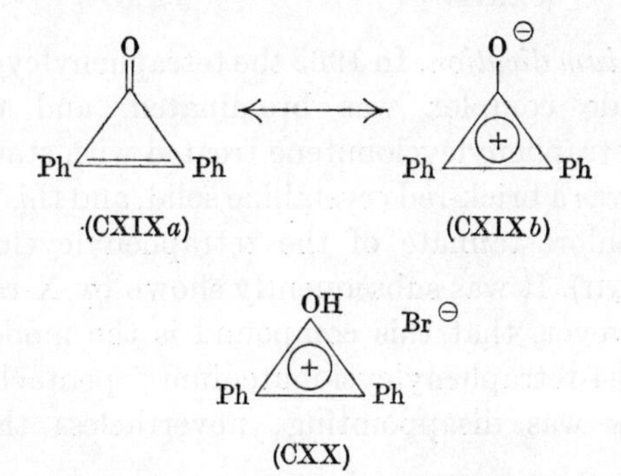

readily forms a stable crystalline hydrobromide (CXX). Other cyclopropenones likewise show high basicity.

Methylenecyclopropene. Methylenecyclopropene (triafulvene) and its derivatives are also of interest, as both classical and dipolar structures can be written. Several simple representatives have been synthesized. The dicyano-derivative (CXXI), for example, has been found to have a high dipole moment ($7{\cdot}9 \pm 0{\cdot}1D$) in agreement with this formulation; and the spectral evidence supports the view that this compound must be regarded as a resonance hybrid. Similarly, the cyano-ethoxycarbonyl derivative has been found to have a dipole moment of $5{\cdot}90 \pm 0{\cdot}1D$.

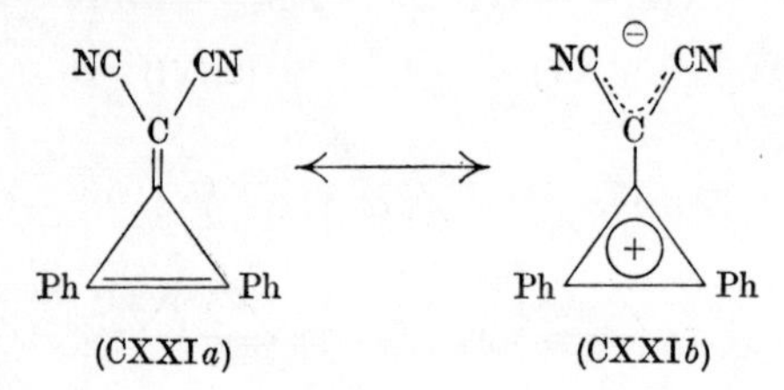

(CXXI*a*) (CXXI*b*)

The cyclopropenium cation also occurs in resonance forms of cyclopropenylidene cyclopentadiene (or calicene) (CXXII). Some derivatives of calicene have been prepared.

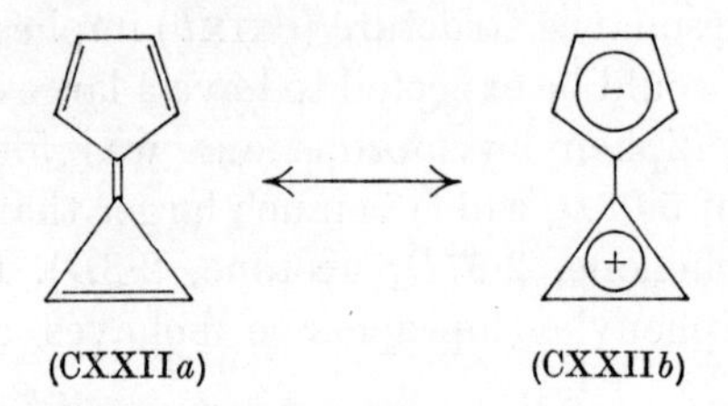

(CXXII*a*) (CXXII*b*)

Cyclobutadienium dication. In 1962 the tetraphenylcyclobutadiene-nickel bromide complex was brominated, and the resulting 3,4-dibromotetraphenylcyclobutene treated with stannic chloride. The product was a brick-red crystalline solid, and this was regarded as the hexachlorostannate of the tetraphenylcyclobutadienium dication (CXXIII). It was subsequently shown by X-ray diffraction methods, however, that this compound is the mono-cation salt, 3-chloro-1,2,3,4-tetraphenylcyclobutenium pentachlorostannate (CXXIV). This was disappointing; nevertheless there is some

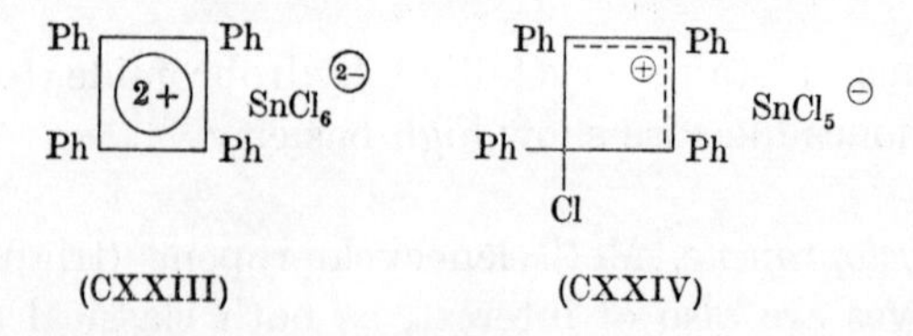

(CXXIII) (CXXIV)

evidence for the existence of a cyclobutadienium dication system. For example, the bromoketone (CXXV) appears to ionize (as shown by n.m.r.) in concentrated sulphuric acid to give the substituted dication (CXXVI).

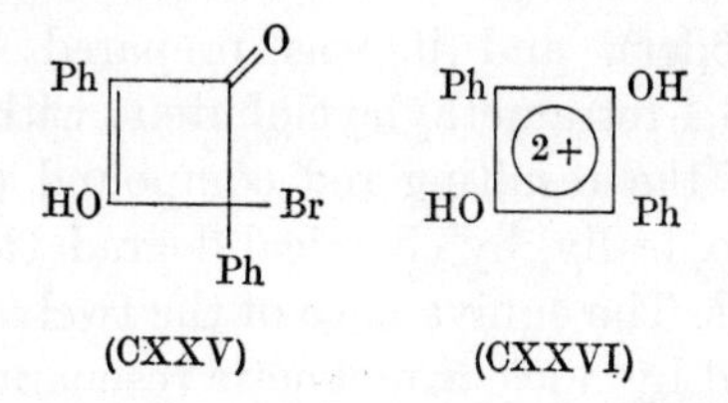

3.3. 4π-Electron systems.

Cyclobutadiene. Cyclobutadiene is the simplest neutral molecule which could conceivably possess aromaticity; but according to Hückel's rule it must be non-aromatic. Many attempts have been made to synthesize cyclobutadiene, and cyclobutadiene derivatives. Perkin attempted to prepare a cyclobutadiene derivative as early as 1894; and Willstätter and von Schmaedel attempted to prepare cyclobutadiene itself in 1905.

If a planar structure is assumed, the molecule of cyclobutadiene would be considerably strained, for the bond angles would need to be 90° instead of the 120° associated with sp^2 by hybridization. Early theoretical studies on butadiene often led to conflicting views on the stability, bond lengths, and other properties; but most investigators agreed that if cyclobutadiene could exist at all it would be unstable and non-aromatic.

Simple Hückel molecular orbital theory predicts a triplet (diradical) ground state (CXXVII) for cyclobutadiene; and Longuet-Higgins & Orgel (1956) further predicted that the diradical would be able to be stabilized as a transition metal complex, $(C_4H_4)MX_2$, M being Ni, Pd or Pt, and X being a univalent element. They further postulated that the nickel complex of cyclobutadiene might be an intermediate in the Reppe synthesis of cyclo-octate-

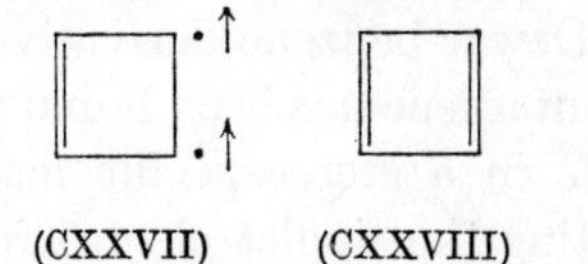

traene. On the other hand, more elaborate calculations lead to the conclusion that cyclobutadiene should have a rectangular arrangement of carbon atoms with alternating, localized, single and double bonds, and having a singlet electronic ground state. According to Dewar & Gleicher (1965*b*), the delocalization energy is virtually zero.

The first complex of cyclobutadiene was reported in 1959

(Criegee & Schröder), and it was prepared by treatment of 3,4-dichloro-1,2,3,4-tetramethylcyclobutene with nickel carbonyl. The structure of the resulting red compound (CXXIX) has been proved spectroscopically, by chemical degradation, and by X-ray diffraction studies. The equivalence of the twelve hydrogen atoms was demonstrated by nuclear magnetic resonance spectroscopy.

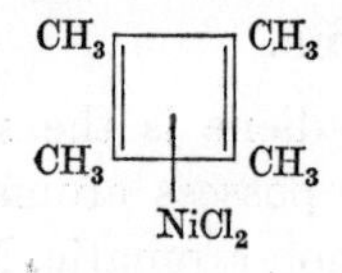

(CXXIX)

The decomposition of the tetramethylcyclobutadiene-nickel(II) complex (CXXIX) leads to octamethyltricyclo-octadiene (the dimer of tetramethylcyclobutadiene) and octamethylcyclo-octatetraene. Many similar complexes of cyclobutadiene derivatives have been prepared. For example, tetraphenylcyclobutadiene complexes, $[C_4Ph_4]X$, have been prepared where X = $Fe(CO)_3$, $NiCl_2$, $NiBr_2$, $PdCl_2$, CoC_5H_5 and $C_4Me_4NiCl_2$.

A stable cyclobutadiene-iron tricarbonyl complex has also been prepared (Emerson, Watts & Pettit, 1965). This complex (CXXX) has been shown to possess aromatic character: it undergoes electrophilic substitution reactions (Fitzpatrick, Watts, Emerson & Pettit, 1965). It has also been found that cyclobutadiene-iron tricarbonyl decomposes in the presence of ceric ions to yield cyclobutadiene. In the absence of other substances this then forms a dimer of cyclobutadiene together with other materials. When the decomposition is conducted in the presence of acetylenic compounds, however, the liberated cyclobutadiene behaves as a diene and reacts to give Dewar-benzene derivatives. In further studies the liberated cyclobutadiene has been found to act both as a diene and as a dienophile in a stereospecific manner. This has been regarded as supporting the singlet electronic ground state structure (CXXVIII) for cyclobutadiene. The n.m.r. spectrum of cyclo-

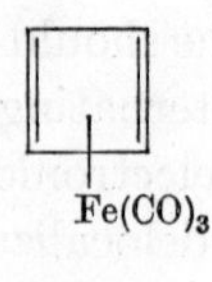

(CXXX)

butadiene-iron tricarbonyl shows a peak at $\tau 6{\cdot}09$, indicating the equivalence of all four protons.

Benzocyclobutadiene. It might be expected that benzocyclobutadiene would be more stable than cyclobutadiene itself, and several attempts have been made to prepare it. 1,2-Dibromobenzocyclobutene (CXXXI) was prepared as early as 1909, but attempts to dehydrobrominate this led to 3-bromo-1,2-benzobiphenylene (CXXXV). It seems that the 1-bromobenzocyclobutadiene (CXXXII) first formed undergoes Diels–Alder dimerization (to CXXXIII) followed by rearrangement (to CXXXIV) and further dehydrobromination (to CXXXV).

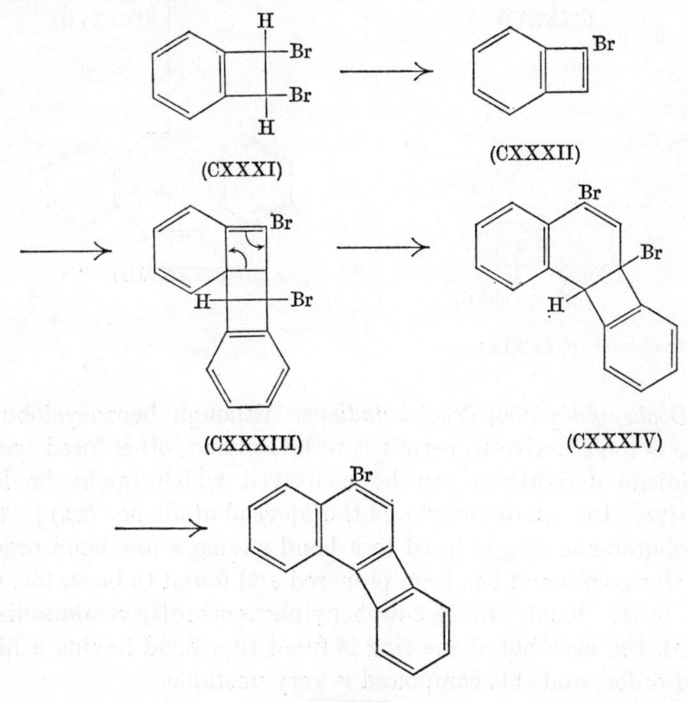

An attempt to trap the intermediate benzocyclobutadiene by heating 1,2-dibromobenzocyclobutane with zinc in furan (hoping thereby to obtain a Diels–Alder adduct with the furan) was unsuccessful; but reaction in the presence of cyclopentadiene instead of furan was successful, the desired adduct being obtained.

The dehydrobromination of 1,2-dibromobenzocyclobutene in a solution containing nickel tetracarbonyl gives a second dimer of benzocyclobutadiene, *viz.* 3,4-7,8-dibenzo-3,7-tricyclo [4.2.0.0^{2,5}] octadiene (CXXXVI). On heating, this is transformed into 1,2-5,6-dibenzocyclo-octatetraene (CXXXVIII). Dehydrobromination of dibromobenzocyclobutene with $Fe_2(CO)_9$ gives benzocyclobutadiene-iron tricarbonyl (CXXXIX), and decomposition of this with silver nitrate yields a dimer of benzocyclobutadiene.

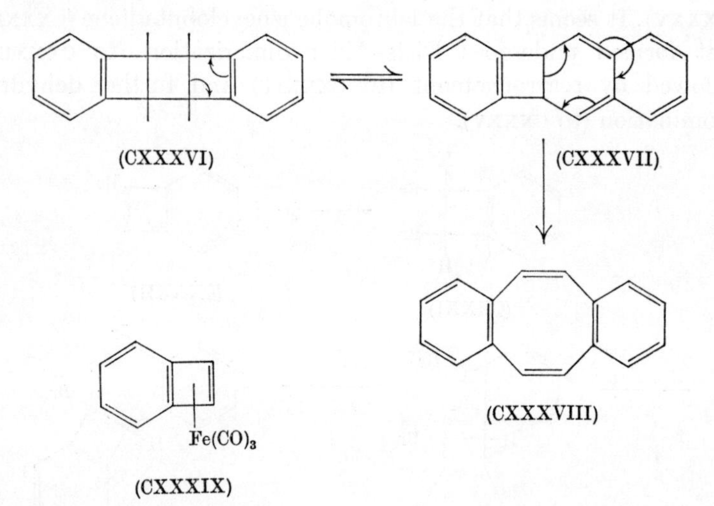

(CXXXVI) (CXXXVII) (CXXXVIII) (CXXXIX)

1,2-*Diphenylnaphtho*[b]*cyclobutadiene.* Although benzocyclobutadiene is too reactive to permit it to be isolated, other fused cyclobutadiene derivatives can be conceived which might be less reactive. In 1,2-diphenylnaphtho[*b*]cyclobutadiene (CXL), the cyclobutadiene ring is fused to a bond having a low bond order; and this compound has been prepared and found to be stable. On the other hand, in 1,2-diphenylphenanthro[*l*]cyclobutadiene (CXLI), the cyclobutadiene ring is fused to a bond having a high bond order; and this compound is very unstable.

Biphenylene. Dibenzocyclobutadiene or biphenylene (CXLII) was first prepared by distillation of 2,2′-dibromo- or 2,2′-diiodobiphenyl with cuprous oxide, and several other methods have since been devised. Of special interest are those involving the intermediate formation of benzyne.

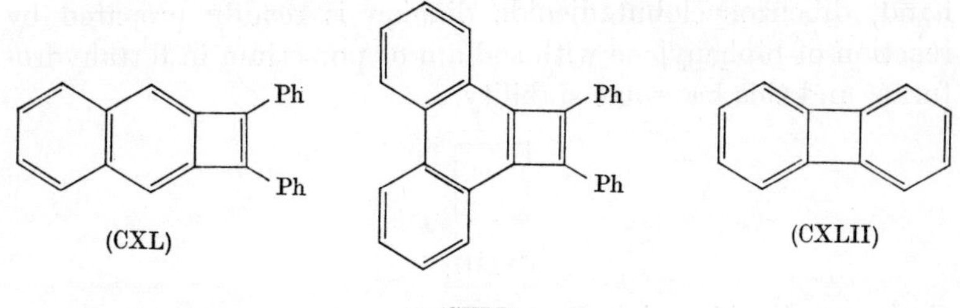

(CXL) (CXLI) (CXLII)

There is no doubt about the structure of biphenylene. It has been confirmed by electron diffraction, and by X-ray crystallographic analysis. It is a stable crystalline compound, and in nearly all respects resembles a typical polycyclic aromatic hydrocarbon (Baker & McOmie, 1959). The ultraviolet absorption spectrum shows two main band systems: the first, of high intensity between 235–260 nm, and the second, of lower intensity, between 330–370 nm. On the other hand, biphenyl has a simple absorption spectrum with an absorption band at 250 nm. Biphenylene undergoes a great variety of substitution reactions, and the chemical evidence, and M.O. calculations, suggest that partial bond fixation occurs in the sense implied by structure (CXLII).

It might be thought that the four-membered ring would be easily opened; but this has not been found to be so. Ring opening is only achieved by reduction reactions. The heat of combustion of crystalline biphenylene has been determined, and this leads to an empirical resonance energy of 17·1 kcal $mole^{-1}$. As the empirical resonance energy of biphenyl is 81·4 kcal $mole^{-1}$, the formation of biphenylene from biphenyl is accompanied by the establishment of strain energy of 64·3 kcal $mole^{-1}$. The n.m.r. spectrum of biphenylene has also been examined, and the two different types of proton have been found to give peaks at τ3·298 and 3·402.

3.4. 6π-Electron systems.

Cyclobutadienide dianion. Hückel's $(4n+2)$ rule leads to the prediction that several 6π-electron systems should possess aromaticity. The simplest is possibly the cyclobutadienide dianion (CXLIII), a system which has not yet been extensively examined. The tetramethyl derivative has, however, been investigated and it has been concluded that it has no special stability. On the other hand, dibenzocyclobutadienide dianion is readily prepared by reaction of biphenylene with sodium or potassium in tetrahydrofuran, and this has some stability.

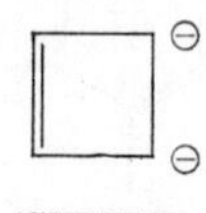

(CXLIII)

Cyclopentadienide anion. Cyclopentadiene is an unsaturated, very reactive hydrocarbon which rapidly undergoes self-condensation (Diels–Alder reaction) to form dicyclopentadiene, from which it can be regenerated by heating. It has a resonance energy of only 2·9 kcal mole^{-1}, in marked contrast to the five-membered heterocyclic compounds pyrrole, furan and thiophen. The acidic nature of cyclopentadiene was recognized by Thiele (1900, 1901), who prepared the cyclopentadienide anion (CXLIV) by the action of potassium in an inert solvent; but a more satisfactory method is to use phenyl lithium (Doering & DePuy, 1953).

Many years ago it was recognized that the cyclopentadienide anion must be regarded as an aromatic ion, having a sextet of electrons, and this has since been confirmed. The anion inflames in air, but this is due to its high reactivity and not to a lack of thermodynamic stability. The symmetrical nature of the anion has been firmly established using carbon-14 as a tracer; and electrophilic substitution reactions have been demonstrated.

The aromaticity of the cyclopentadienide anion is confirmed by the n.m.r. spectrum, which shows a peak at τ4·66. This establishes the presence of a ring current in the negatively charged ring.

Derivatives of the cyclopentadienide anion are known. The tetracyano derivative (CXLV) is stable, and undergoes electrophilic substitution under mild conditions. It has been nitrated, brominated and acylated.

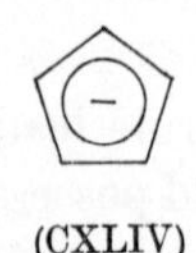

(CXLIV)

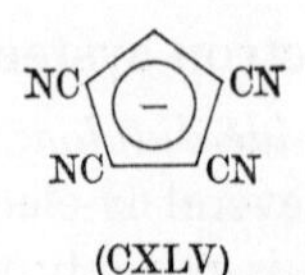

(CXLV)

Diazocyclopentadiene and related compounds. The negatively charged cyclopentadienide ring system is also found in many dipolar molecules such that an exocyclic positive charge balances the negative charge distributed over the five-membered ring. The first such *ylide* to be prepared was diazocyclopentadiene (CXLVI) (Doering & DePuy, 1953). It is a red solid, m.p. −23 to −22°, which decomposes on standing at 0°, and there have been reports of spontaneous explosive decomposition; but it has been shown to undergo a variety of electrophilic substitution reactions, including nitration, diazo-coupling and bromination. The ultraviolet absorption spectrum of diazocyclopentadiene shows a maximum at 298 nm and a shoulder at ~340 nm, with long tailing absorption into the visible region, which is responsible for the colour. There can be no doubt that this spectrum results from a highly conjugated system; and it may also be noted that substituted diazocyclopentadienes absorb at longer wavelengths.

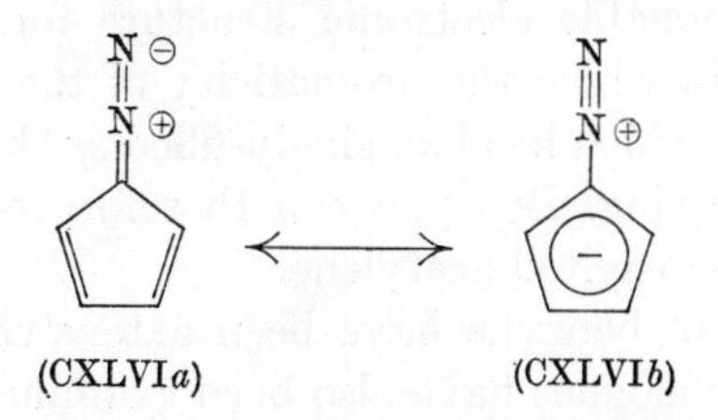

(CXLVI*a*) (CXLVI*b*)

The n.m.r. spectrum of diazocyclopentadiene confirms the aromaticity; it shows two multiplets of equal area centred at τ3·3 and 4·2; the τ3·3 bands have been assigned to the 3,4-hydrogens and the τ4·2 bands to the 2,5-hydrogens. The cyclopentadienide anion, which carries a full negative charge in the ring, absorbs at τ4·66, not far from the 3,4-hydrogens in diazocyclopentadiene.

Benzene and benzyne. The most important 6π-electron system is, of course, benzene. The reactions of this hydrocarbon exemplify what is usually understood by 'aromatic character'; and the electronic structure of benzene exemplifies what is referred to as 'aromaticity' (see chapter 1).

The abstraction of a hydrogen atom from a benzene molecule gives a phenyl radical. Benzyne, or dehydrobenzene, results from the abstraction of two hydrogen atoms from *ortho* positions in a benzene molecule. Considerable evidence has now accumulated that benzyne is formed as an intermediate in a number of reactions, and several methods for the generation of benzyne have been devised (see Heaney, 1962). Among these may be mentioned the reaction of halobenzenes with aryl-lithium or alkyl-lithium, for example the reaction of *o*-bromohalobenzene with butyl-lithium. Another method involves the decomposition of benzene-diazonium-2-carboxylate; and a third convenient method involves the oxidation of 1-aminobenzotriazole. Evidence for gaseous benzyne has been obtained from the flash photolysis of the diazonium salt.

Benzyne is so tremendously reactive that it is generally necessary to infer its existence as an intermediate by the nature and course of the reaction. Nevertheless, the existence of benzyne as a free molecular species has been reported following mass spectrometrical experiments.

The most reasonable electronic structure for benzyne would seem to be one in which the aromaticity of the cyclic system is undisturbed, but which has two singly-filled sp^2 hybrid orbitals in the plane of the ring. Benzyne can therefore be regarded as a diradical, or as an excited acetylene.

The reactions of benzyne have been extensively studied, and many benzyne analogues have also been examined. For example, evidence for the intermediate existence of 9,10-dehydrophenanthrene, and of heterocyclic analogues (known as hetarynes) has been provided.

Cycloheptatrienium cation. The cycloheptatriene (tropilidene) molecule has one saturated carbon atom and cannot therefore be fully conjugated. The methylene group is tilted away from the plane of the adjacent double bonds, and the molecule is subject to ring inversion. Cycloheptatriene behaves as an olefin, and its empirical resonance energy, calculated from the heat of hydrogenation, has been found to be 6·7 kcal mole^{-1} in agreement with this formulation.

On the other hand, the cycloheptatrienium cation (tropylium or tropenium cation) (CXLVII) has six π-electrons, and according to Hückel's rule should possess aromaticity. The cycloheptatrienium or tropylium cation was first obtained in 1891, but it was not recognized as such until later (Doering & Knox, 1954). Tropylium bromide was found to be a strongly deliquescent solid; and tropylium isocyanate has also been obtained, as have other salts.

It is of interest that the tropylium cation can itself behave as an electrophilic reagent; it reacts with 2,6-dimethylphenol, for example, to yield a substituted tropylium cation.

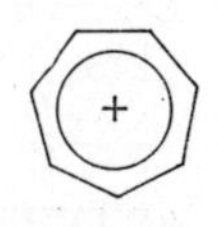

(CXLVII)

The symmetrical nature of the tropylium ion has been established beyond doubt. Labelled tropylium bromide was prepared from benzene and $^{14}CH_2N_2$, and then treated with phenylmagnesium bromide to give phenylcycloheptatriene. Oxidation of this gave benzoic acid with radioactivity corresponding to one-seventh of a labelled carbon atom. The phenylmagnesium bromide must, therefore, have reacted equally with all seven ring carbon atoms.

Attempts to obtain a detailed structure of the tropylium ion by X-ray crystallography have been unsuccessful owing to the disorder in the crystals. The spectral evidence, however, confirms the aromaticity of the ion. The ultraviolet absorption spectrum shows maxima at 247 and 275 nm with long tailing into the visible region. The n.m.r. spectrum consists of a single line at $\tau 0{\cdot}91$, confirming the existence of a ring current in the positively charged ring system.

Derivatives of the tropylium cation are of special interest; in some of these the aromatic character is enhanced. First among these may be mentioned cycloheptatrienone (=tropone), and its hydroxy-derivatives.

Tropone (CXLVIII) is a resonance hybrid of the classical structure (CXLVIII*a*) and the tropylium cation structure (CXLVIII*b*). It is readily hydrogenated and rapidly decolorizes aqueous permanganate, but it gives no 2,4-dinitrophenylhydrazone, and reacts with hydroxylamine only on heating. The contribution of the ionic structure is confirmed by the dipole moment ($4{\cdot}17D$) which is relatively high compared with that of benzophenone or acetone. It is also confirmed by the fact that tropone readily adds a proton to give the hydroxytropylium cation (CXLIX). Nevertheless this contribution cannot be very great as tropone displays only limited aromatic character; and X-ray structure determination of 2-chlorotropone shows bond-length alternation in the sense predicted by the classical structure (CXLVIII*a*). Nevertheless, the empirical resonance energy of tropone, from heats of combustion, is ~ 29 kcal mole^{-1}. The ultraviolet spectrum of tropone (in iso-

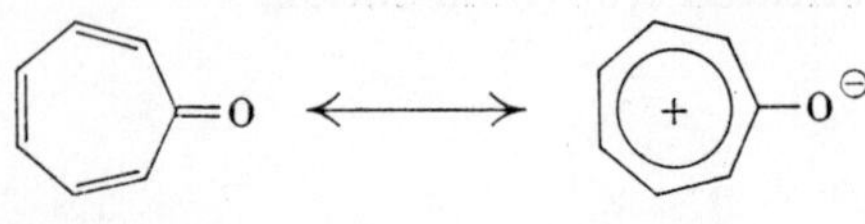

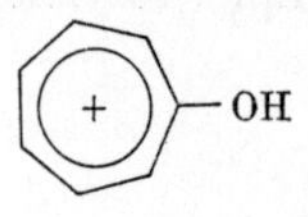

(CXLVIII*a*) 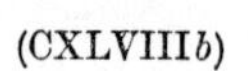 (CXLVIII*b*) (CXLIX)

Bond lengths in 2-chlorotropone

octane) shows maxima at 225, 297 and 310 nm; and the infrared spectrum shows C—H stretching bands at 3060–3010 cm^{-1}, similar to benzene. Tropone gives a carbonyl band at 1638 cm^{-1}, a lower wave number than for normal ketones.

2,3-Benzotropone and 2,3-6,7-dibenzotropone have also been prepared, and in these systems the tropone ring exhibits even less aromaticity. 2,3-Benzotropone is less polar (3·61*D*) than tropone. Dibenzotropone is much less basic than tropone, and the carbonyl frequency has the same value as that for benzophenone (1660 cm^{-1}).

Much greater stability and aromatic character are displayed by 2-hydroxytropone, or tropolone (CL). The proximity of the hydroxy-group to the carbonyl group facilitates intramolecular hydrogen bonding; this is confirmed by a shift in the OH absorption in the infrared from the normal value (phenol, 3600 cm^{-1}) to 3100 cm^{-1}. Moreover, the proximity of the two groups also leads to tautomerism, and resonance structures may be written for the tautomer (CLI).

The tropolone ring system occurs in several natural products, including stipitatic acid, puberulic acid, thujaplicins and colchicine. Indeed, a tropolone ring system was first postulated to explain the properties of stipitatic acid (Dewar, 1945). Since then the tropolone ring system has been extensively studied (see Pauson, 1955*a*; Nozoe, 1959), and there is no doubt that it possesses marked aromatic character. Tropolone is not hydro-

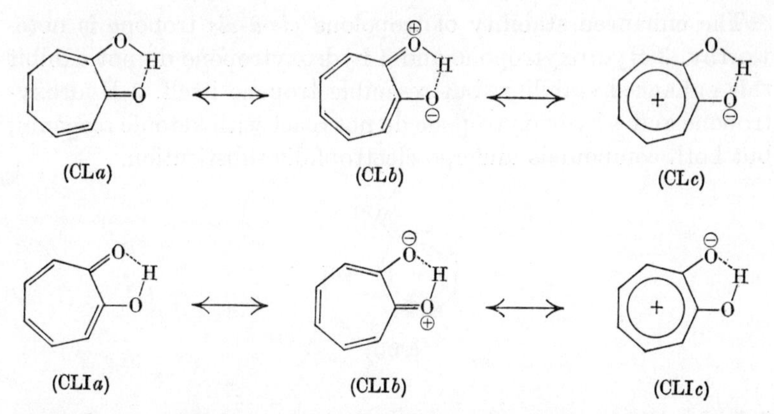

genated over palladium, but over platinum it is hydrogenated to a mixture of *cis*- and *trans*-cycloheptane-1,2-diols. In many ways it behaves like a phenol. It forms metallic salts, ethers, and esters; and it has pk_a 7·00 (compare 10·0 for phenol). It shows strong absorption in the ultraviolet, which changes in going from neutral to alkaline solution in a manner similar to that observed with phenols. In the infrared the carbonyl stretching frequency is abnormally low (1615 cm^{-1}). Tropolone undergoes a wide variety of electrophilic substitution reactions.

The aromatic nature of tropolone has been confirmed by X-ray structure analysis of cupric tropolone. It has an almost regular, planar heptagonal structure, with an average carbon–carbon bond length of 1·40 Å. The two carbon–oxygen bond lengths were found to be different; the carbonyl oxygen was 1·25 Å from the ring; the other oxygen was 1·34 Å from the ring. Moreover, the carbonyl oxygen was 1·98 Å from the copper atom, and the other, 1·83 Å. It is also of interest that the theoretical angle for a regular heptagon is 128·6°, and that the observed values were all within 4° of this. Other structural investigations by physical methods include electron diffraction studies on tropolone, and X-ray studies of the crystal structure of the hydrochloride, and the sodium salt. In the hydrochloride both C—O distances were found to be the same, 1·36 Å, in agreement with the symmetrical structure (CLII).

The empirical resonance energy of tropolone, from thermochemical data, is 36 kcal $mole^{-1}$; and finally, the ultraviolet spectrum (in cyclohexane) shows maxima at 222, 232, 238, 322, 340, 356 and 374 nm.

The enhanced stability of tropolone *vis-à-vis* tropone is noteworthy. 3-Hydroxytropone and 4-hydroxytropone do not exhibit this enhanced stability, but resemble tropone itself. 3-Hydroxytropone and 4-hydroxytropone do not react with ketonic reagents; but both compounds undergo electrophilic substitution.

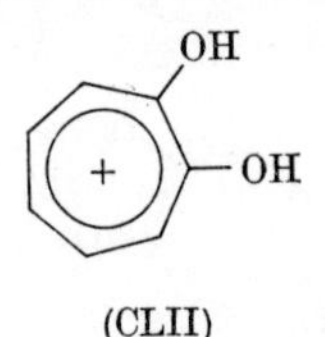

(CLII)

3.5. 8π-Electron systems.

Cyclo-octatetraene. Cyclo-octatetraene (CLIII) was first prepared in 1911 by Willstätter and Waser by a thirteen-step series of degradations from the pomegranate alkaloid, pseudopelletierine. Willstätter's hydrocarbon was found to be extremely reactive; and some years later doubts were expressed about its structure. In particular, attention was directed to the striking similarity between Willstätter's hydrocarbon and styrene. Subsequent work has, however, shown that degradation of pseudopelletierine does indeed give cyclo-octatetraene (CLIII). Willstätter's original work has been repeated and confirmed.

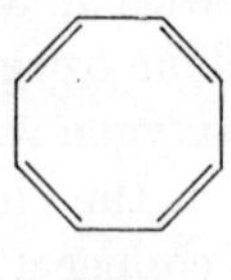

(CLIII)

In the meantime, a synthesis of cyclo-octatetraene had been developed by Reppe in the laboratories of the I.G. Farbenindustrie at Ludwigshafen; but the results were not published until the end of the war. This synthesis involved the polymerization of acetylene in tetrahydrofuran solution in the presence of nickel cyanide and ethylene oxide at a pressure of 15–20 atmospheres and 60–70°. Cyclo-octatetraene was isolated in about 70 % yield, together with some benzene and smaller amounts of other hydrocarbons (naphthalene, azulene, vinylcyclo-octatetraene and 1-phenyl-1,3-butadiene). Cyclo-octatetraene prepared in this way was found to be identical with that prepared by degradation from pseudo-pelletierine.

The mechanism of the Reppe synthesis has attracted considerable attention, especially since the suggestion that a cyclobutadiene-nickel cyanide complex may be involved. A bimolecular reaction between two C_4H_4 units would explain the predominance of cyclo-octatetraene over the thermodynamically more stable benzene.

Cyclo-octatetraene, having 8π-electrons, does not conform to Hückel's rule which therefore predicts that the hydrocarbon should be non-aromatic. The chemistry of cyclo-octatetraene has been extensively studied (see Raphael, 1959), and there is no doubt that cyclo-octatetraene is a cyclic polyene, and not an aromatic compound. It is a very reactive substance. It undergoes ready catalytic hydrogenation to cyclo-octene and to cyclo-octane. It is oxidized to aromatic compounds such as benzaldehyde, benzoic acid, and phthalic acid. With per-acids it gives an epoxide, 5,8-epoxy-1,3,6-cyclo-octatriene. In the original work it was shown that chlorine, bromine and hydrogen bromide react with cyclo-octatetraene by addition; but the structures of the compounds were not established. It is now known that these reactions give derivatives of bicyclo[4.2.0]octane. Chlorine, for example, gives 7,8-dichloro-2,4-bicyclo[4.2.0]octadiene (CLIV).

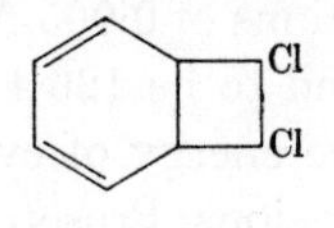

(CLIV)

In the Diels–Alder reaction, cyclo-octatetraene reacts with one molecule of dienophile. The reaction proceeds very readily, so readily that cyclo-octatetraene also undergoes the Diels–Alder reaction with itself as a dienophile. Two crystalline dimers have been isolated at temperatures below 100°, and above 100° two different isomers have been obtained. Kinetic studies of the Diels–Alder reaction have shown that there is a time-valence tautomerism between cyclo-octatetraene (CLIII) and 2,4,7-bicyclo[4.2.0]octatriene (CLV). It may reasonably be concluded that cyclo-octatetraene is a very reactive and relatively unstable ring system, quite unlike benzene in its properties.

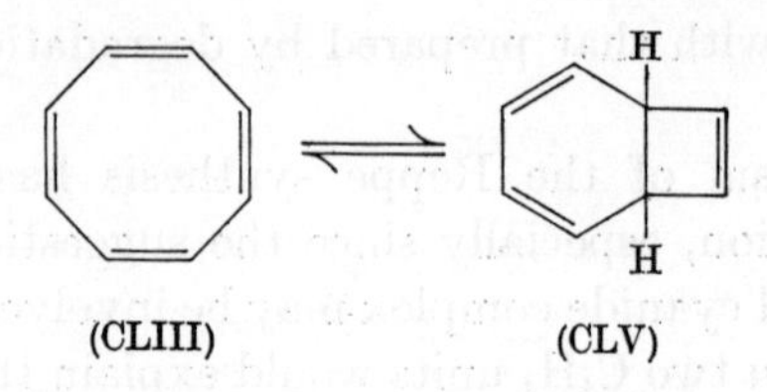

For aromatic stability and high resonance energy the molecule would have to be planar; in such a case, however, the angular valency angles would be 135° as against 120° for sp^2 hybridization characteristic of aromatic compounds. The experimental evidence indicates that the molecule is not planar, and that the bonds are not aromatic, but approximate to alternate single and double bonds. The weight of evidence suggests that cyclo-octatetraene has a puckered D_{2d} 'tub' conformation. From an electron diffraction study it seems that the C—C bonds are 1·50 Å, the C=C bonds 1·35 Å, the C—H bonds 1·13 Å, and the C=C—H angle 124°. Moreover the Raman spectrum shows a strong band indicative of isolated C=C bonds. An X-ray analysis indicates that the carbon–carbon bonds are alternately 1·34 and 1·54 Å and the C=C—C angles 125°. Similarly X-ray analysis of the crystal structure of cyclo-octatetraenecarboxylic acid has established a 'tub' configuration with alternating single and double bonds. The mean bond lengths were found to be 1·470 and 1·322 Å respectively, with estimated standard deviations of 0·005 Å; and the mean interbond angle in the ring was found to be $126{\cdot}4 \pm 0{\cdot}4°$.

The empirical resonance energy of cyclo-octatetraene has been determined on several occasions. Prosen, Johnson & Rossini (1950) determined the heat of combustion and concluded that cyclo-octatetraene is less stable than styrene by 36·1 kcal mole^{-1}. The heat of combustion has also been determined by Springall and co-workers (1954). Their value (which is in good agreement with the earlier one) leads to an empirical resonance energy of 4·8 kcal mole^{-1} for cyclo-octatetraene. From the heat of hydrogenation, an empirical resonance energy of 2·4 kcal mole^{-1} has been found.

The ultraviolet absorption spectrum of cyclo-octatetraene (in cyclohexane) shows a maximum at 282 nm (log ϵ, 2·4); and the n.m.r. spectrum shows a peak at τ4·31, confirming the absence of an induced ring current. It seems certain, therefore, that cyclo-octatetraene is non-planar and non-aromatic.

Dibenzocyclo-octatetraene. This compound (CLVI) has been prepared in several ways. It decolorizes permanganate solution, and bromine in carbon tetrachloride. The reaction with bromine involves normal addition to the olefinic linkages; and both a highly reactive dibromide and a very unreactive tetrabromide can be obtained.

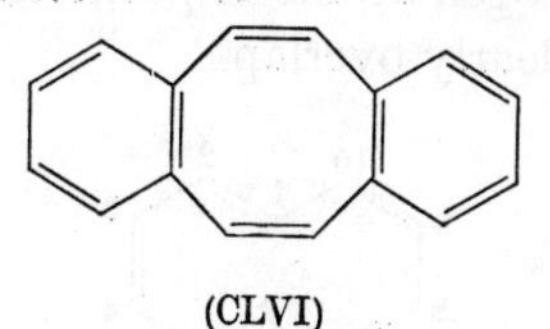

(CLVI)

Tetrabenzocyclo-octatetraene. This compound, also known as tetraphenylene (CLVII), is a colourless high-melting stable compound. It is recovered unchanged after treatment with potassium permanganate in boiling acetone. The central cyclo-octatetraene ring has the 'tub' conformation, and the four benzene rings are directed towards the corners of a regular tetrahedron.

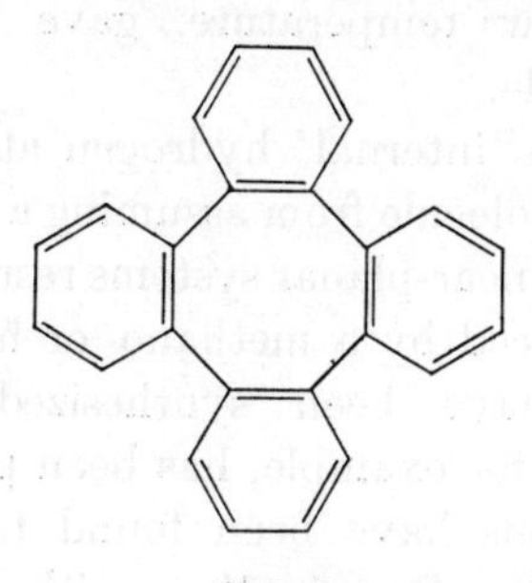

(CLVII)

Cycloheptatrienide anion. This anion (CLVIII) also represents a potential 8π-electron system. It has been prepared as the potassium salt; and experiments on the base-catalyzed deuterium exchange in cycloheptatriene also provide evidence for the existence of the cycloheptatrienide anion.

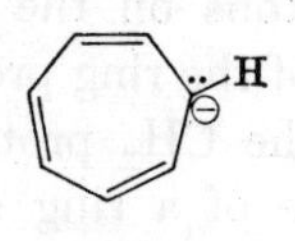

(CLVIII)

3.6. 10π-Electron systems.

[10]*Annulene.* [10]Annulene, or 1,3,5,7,9-cyclodecapentaene(CLIX) is the next possible $(4n+2)$ neutral analogue of benzene: for benzene, $n = 1$; for [10]annulene, $n = 2$. However, high resonance stabilization would seem to be unlikely as the van der Waals radii of the 'internal' hydrogen atoms at positions 1 and 6 on the *trans* double bonds must clearly overlap.

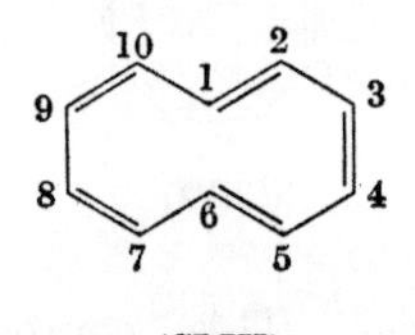

(CLIX)

Attempts have been made to prepare [10]annulene; one involved the preparation of *cis*-9,10-dihydronaphthalene; but this was found to be stable at room temperature and did not isomerize to [10]annulene. On the other hand, *trans*-9,10-dihydronaphthalene, on irradiation at a low temperature, gave [10]annulene, which proved to be unstable.

Although the two 'internal' hydrogen atoms in [10]annulene would prevent the molecule from assuming a planar configuration, models indicate that near-planar systems result if these two hydrogen atoms are replaced by a methano- or hetero-bridge. Several such compounds have been synthesized. 1,6-Methanocyclodecapentaene (CLX), for example, has been prepared and studied. Substitution reactions have been found to give mono- or disubstituted products. Bromination with *N*-bromosuccinimide gives a yellow monobromide in 90 % yield; and excess *N*-bromosuccinimide gives a dibromo-derivative. Acetylation with acetic anhydride gives a monoacetyl derivative. The ultraviolet absorption spectrum shows maxima at 256, 259 and 298 nm confirming the extended conjugation. The n.m.r. spectrum shows an A_2B_2 system at $\tau 2{\cdot}5$ to $3{\cdot}2$, centred at $\tau 2{\cdot}8$; and a sharp signal at $\tau 10{\cdot}5$, assigned to the two protons on the methylene bridge, is also observed. The resonance of the ring protons at very low field, and the strong shielding of the CH_2 protons, can be considered as evidence for the presence of a ring current. 1,6-Methanocyclodecapentaene may therefore be represented as in (CLXI).

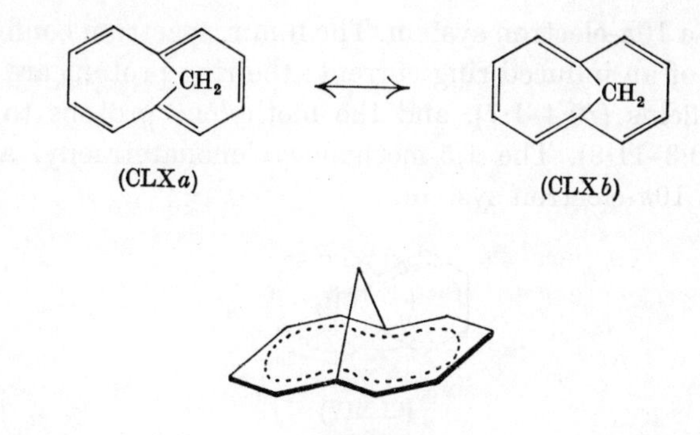

(CLXa) (CLXb)

(CLXI)

This conclusion has been supported by X-ray structure analysis of the 2-carboxylic acid. This compound was found to have an approximately planar C_{10} perimeter, and the lengths of the carbon–carbon bonds were found not to show any appreciable alternation. A tricarbonyl-chromium derivative of methanocyclodecapentaene has also been prepared.

The chemistry of 1,6-methano[10]annulene has been investigated. Many substitution derivatives have been prepared and a variety of transformations of substituents has been effected. A benzyne-type intermediate has also been postulated.

1,6-Oxido [10]annulene (CLXII) and 1,6-imino [10]annulene (CLXIII) have also been described. The former undergoes nitration, giving two isomeric mononitro derivatives. The ultraviolet spectrum showed maxima at 255, and 299 nm, and a complex band at *ca.* 392 nm, and generally resembled that given by the 1,6-methano derivative. The n.m.r. spectrum showed an A_2B_2 pattern in the $\tau 2{\cdot}23$ to $2{\cdot}81$ region, centred at $\tau 2{\cdot}52$, similar to that of naphthalene ($\tau 2{\cdot}05$ to $2{\cdot}72$, centred at $\tau 2{\cdot}38$). The spectra of the 1,6-imino compound were found to be similar.

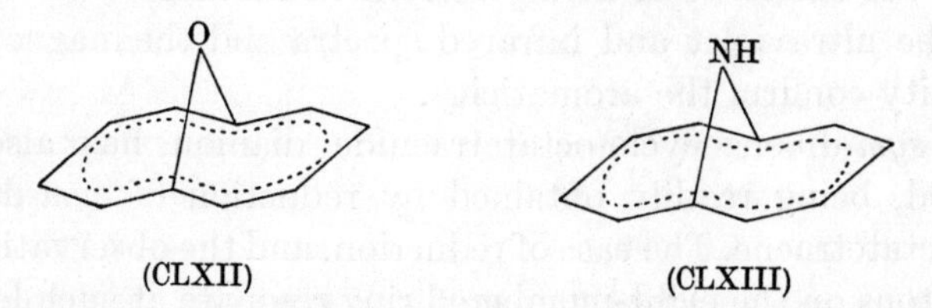

(CLXII) (CLXIII)

It is of some interest that bridged ionic systems have also been reported. Thus the bicyclo[5.4.1]dodecapentaenium cation

(CLXIV) is a 10π-electron system. The n.m.r. spectrum confirms the existence of an induced ring current: the ring protons are shifted to lower fields (τ0·4–1·7), and the methylene protons to higher fields (τ10·3–11·8). The 1,5-methanocyclononatetraenyl anion is likewise a 10π-electron system.

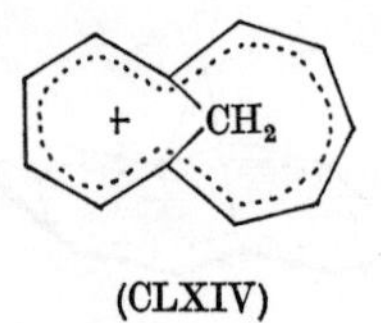

(CLXIV)

Cyclononatetraenide anion. This anion (CLXV) has 10π-electrons, and the spectra indicate that it must be considered an aromatic system. The ultraviolet spectrum of the anion, in tetrahydrofuran,

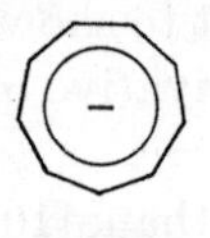

(CLXV)

showed maxima at 252, 318 and 325 nm. The n.m.r. spectra of potassium and of lithium cyclononatetraenide in completely deuterated tetrahydrofuran showed a single peak at τ2·96 and 3·15 respectively; this may be compared with the peak at τ4·3 found for the protons of cyclo-octatetraenide dianion. Studies of the ^{13}C n.m.r. spectrum also support the conclusion.

Cyclo-octatetraenide dianion. This dianion also has 10π-electrons and unlike the neutral molecule is planar and aromatic. In the n.m.r. spectrum, for example, the protons absorb at τ4·3, confirming the existence of a ring current in the negatively charged ring. The ultraviolet and infrared spectra and the magnetic susceptibility confirm the aromaticity.

The *sym*-dibenzocyclo-octatetraenide dianion has also been prepared, being readily obtained by reduction of *sym*-dibenzocyclo-octatetraene. The ease of reduction, and the observation that the protons on the eight-membered ring resonate at such low field (τ2·82) in the n.m.r. spectrum, suggest that this dianion is also aromatic.

Cycl[3.2.2.]azine. Attention may also be directed to a bridged [10]annulene in which a nitrogen atom acts as a link to three carbon atoms. This is the tricyclic cycl[3.2.2]azine (CLXVI). It is a stable crystalline compound with an odour indistinguishable from that of naphthalene; and its properties are typical of an aromatic compound. It forms a *sym*-trinitrobenzene complex, and undergoes electrophilic substitution reactions smoothly and in good yield. The ultraviolet absorption spectrum shows three groups of absorption bands. The Group I band has a maximum at 244 nm (log ϵ, 4·57); the Group II bands at 274 and 289 nm (log ϵ, 3·74, 3·86); and the Group III bands at 398, 408 and 419 nm (log ϵ, 3·56, 3·66 and 3·65). The n.m.r. spectrum of cyclazine shows peaks with τ values assigned to the protons as indicated in (CLXVI). A ring current may be inferred from these data and the aromaticity of cyclazine is established.

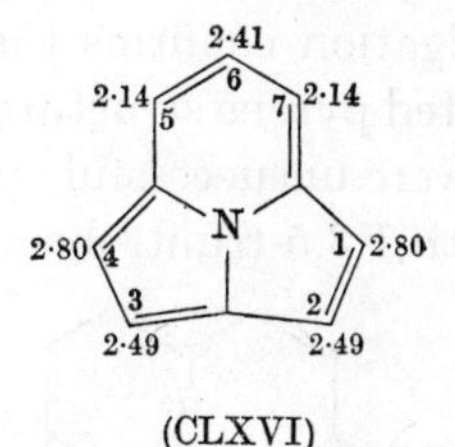

(CLXVI)

3.7. 12π-Electron systems.

[12]Annulene. This compound does not conform to Hückel's rule and preliminary work suggests that it is an unstable, reactive hydrocarbon. Two isomeric bisdehydro[12]annulenes have been described, but one has since been found to be tridehydro[12]annulene (CLXVII).

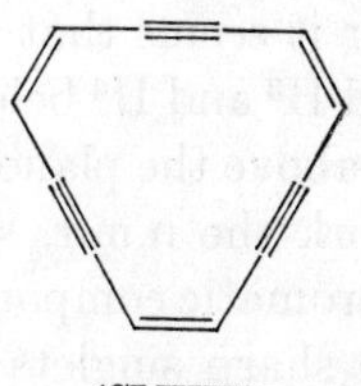

(CLXVII)

This interesting hydrocarbon (CLXVII) contains $4n$ π-electrons in parallel p orbitals, and there is every reason to expect it to be planar and strainless. It does not conform to Hückel's rule and it was therefore of great importance to examine it for aromaticity. It is a dark brick-red, crystalline compound which reacted readily with oxygen, both in solution and in the solid state. The ultraviolet spectrum of this compound in iso-octane showed maxima at 239 nm (ϵ, 78,700) and 247·5 nm (ϵ, 57,200), and low intensity absorption throughout the visible range with a maximum at 457 nm (ϵ, 177). The n.m.r. spectrum showed a single sharp absorption peak at τ5·55. It must be concluded therefore that tridehydro[12]annulene is not aromatic.

3.8. 14π-Electron systems.

[14]Annulene. Cyclotetradecaheptaene or [14]annulene (CLXVIII) conforms to Hückel's rule, but models and scale drawings indicate that the four 'internal' hydrogen atoms overlap so that a planar molecule is unlikely. [14]Annulene has been prepared and X-ray crystallographic investigation confirms that the molecule corresponds to a highly distorted pyrene structure. Attempts to carry out substitution reactions were unsuccessful, and the hydrocarbon did not form an adduct with 1,3,5-trinitrobenzene.

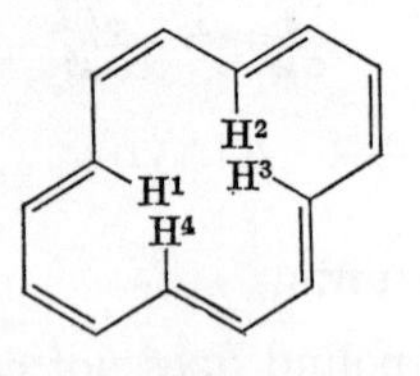

(CLXVIII)

The overcrowding of the four internal hydrogen atoms in [14]annulene is responsible for the existence of two conformational isomers, and these have been separated by thin-layer chromatography. In one conformer it seems that H^1 and H^2 are above the plane of the molecule, and H^3 and H^4 below the plane; in the other conformer H^1 and H^3 are above the plane and H^2 and H^4 below the plane. In preliminary work the n.m.r. spectrum of [14]annulene seemed to suggest a non-aromatic compound. At room temperature the spectrum shows two sharp singlets at τ4·42 and 3·93 (ratio *ca.* 6:1) due to the two conformers. The τ4·4 band (isomer A) has been shown to undergo fundamental changes when the solution is cooled. It broadens progressively, and with further cooling eventually disappears, being replaced by a very broad new band at *ca.* τ2·7. At −60° the spectrum shows broad peaks at τ2·4 and at τ10·0 due to the outer and inner protons respectively of isomer A and a peak at τ3·9 due to isomer B. This low temperature spectrum of isomer A is typical of an aromatic substance as it suggests the existence of a ring current which deshields the outer protons and shields the inner protons. It is possible that [14]annulene can sustain an induced ring current even at room temperature, and that the singlets at τ4·42 and 3·93 may be averages for the chemical shifts of the inner and outer protons due to rapid interchanges of proton positions.

The overlapping of the van der Waals radii of the four inner hydrogen atoms in [14]annulene clearly reduces the aromaticity of this compound. With monodehydro[14]annulene (CLXIX) and with 1,8-bisdehydro[14]annulene (CLXX), however, no such overlapping occurs and models of these compounds suggest planar configurations.

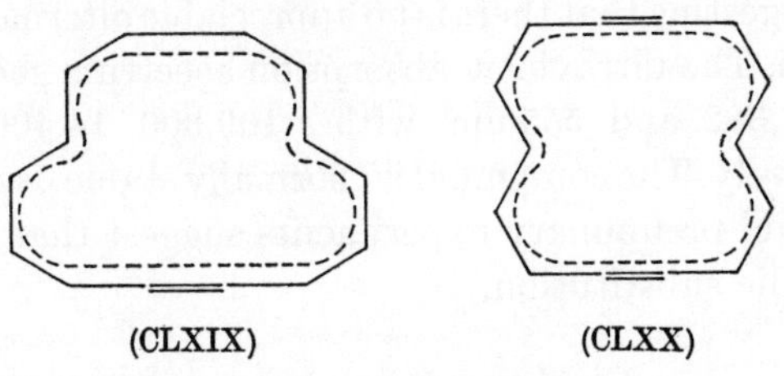

(CLXIX) (CLXX)

There is no doubt that 1,8-bisdehydro[14]annulene is an aromatic compound. It forms a 1:1 adduct with 1,3,5-trinitrobenzene, and it readily undergoes electrophilic substitution reactions. The n.m.r. spectrum in deuterochloroform shows three multiplets at τ15·54, 1·57 and 0·45, with relative intensities 2:4:4. The band at τ15·54 may be assigned to the internal protons, and those at τ0·45 and 1·57 to the external protons. The ultraviolet absorption spectrum of 1,8-bisdehydro[14]annulene in iso-octane showed a maximum at 586 nm; and determination of the X-ray crystal structure of this hydrocarbon showed that the carbon–carbon bond lengths are 1·374 to 1·400 Å, with 1·200 Å for the classical triple bond.

Monodehydro[14]annulene (CLXIX) is similarly aromatic. It also forms a 1:1 adduct with 1,3,5-trinitrobenzene, and it undergoes electrophilic substitution reactions. The n.m.r. spectrum shows a complex band in the region τ1·2–2·7 (outer protons) and a double doublet at τ10·70 (inner protons), confirming the existence of a ring current.

A bridged [14]annulene has been reported. This compound (CLXXII) has been prepared from the bis-epoxide of 1,4,5,8,9,10-hexahydroanthracene (CLXXI) by bromination followed by dehydrobromination. It has been found to be a carmine-red crystalline substance. Its n.m.r. spectrum shows a singlet at τ2·06 due

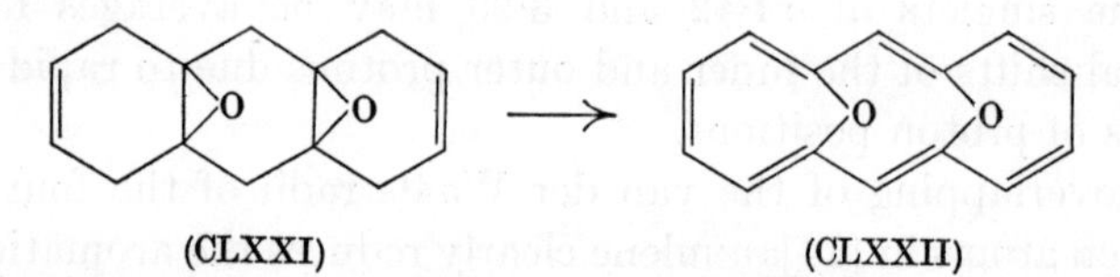

(CLXXI) (CLXXII)

to the two *meso* protons and an AA′BB′ system (τ_A = 2·25, τ_B = 2·40) arising from the remaining protons. The coupling constants, J_{AB} = 9·0 and J_{BB} = 9·2, for vicinal protons are almost identical suggesting that there is no appreciable alternation in C—C bond lengths. The ultraviolet absorption spectrum shows maxima at 306, 345, 382 and 555 nm, with ϵ 169,000, 14,400, 8,500 and 775 respectively. The compound is thermally stable, is not sensitive to oxygen, and preliminary experiments suggest that it is subject to electrophilic substitution.

15,16-Dihydropyrene. [14]Annulene may be considered as an open analogue of 15,16-dihydropyrene (CLXXIII), and derivatives of the latter have been prepared. An example is 2,7-diacetoxy-*trans*-15,16-dimethyl-15,16-dihydropyrene (CLXXIV); its n.m.r. spectrum showed peaks at τ1·42 and 1·63 due to the external protons, at τ14·03 due to the internal methyl groups, and at τ7·50 due to the acetyl groups. This spectrum clearly establishes the existence of a ring current, and the aromaticity of this interesting compound is further confirmed by the ultraviolet-visible absorption spectrum.

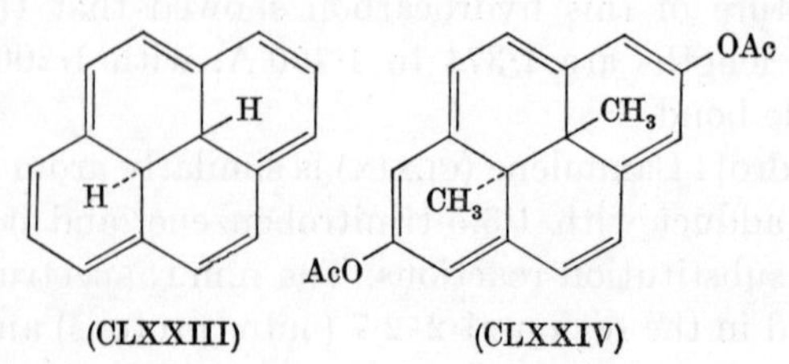

(CLXXIII) (CLXXIV)

The simpler *trans*-15,16-dimethyl-15,16-dihydropyrene (CLXXIV without the AcO-groups) is an emerald-green crystalline compound, and its absorption spectrum showed maxima at 337·5, 377, 463, 528, 536, 586, 598, 610, 627, 634 and 641 nm. The n.m.r. spectrum showed signals at τ1·33 (6 protons), τ1·43 (2 protons), a multiplet at τ1·77–2·02 (2 protons), and a sharp singlet at τ14·25 (6 protons). The displacement of the ring protons to low field, and the remarkable shift of the protons of the internal methyl groups to high field, provide clear evidence for the existence of a ring current. In addition, however, this compound has been found to undergo a variety of electrophilic substitution reactions.

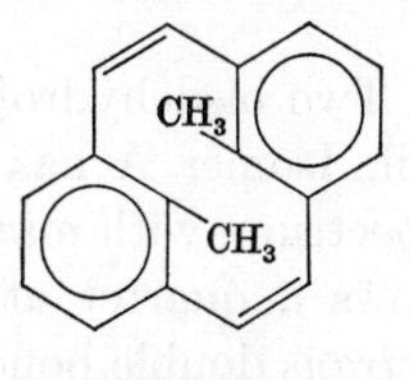

(CLXXV)

It is also of interest that this hydrocarbon undergoes photoisomerization to its valence tautomer (CLXXV) having a metacyclophane structure. In the dark the metacyclophane reverts to the more stable dihydropyrene.

3.9. 16π-Electron systems.

[16]Annulene. This compound, having 16π-electrons, does not conform to Hückel's rule, and should therefore be non-aromatic. It has been prepared by partial hydrogenation of bisdehydro[16]-annulene and by irradiation of a cyclo-octatetraene dimer, and its physical properties confirm its non-aromatic nature.

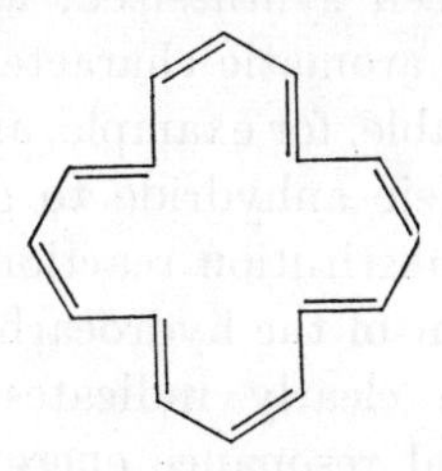

(CLXXVI)

Solutions of [16]annulene are deep red: and the ultraviolet absorption spectrum of [16]annulene in cyclohexane shows maxima at 282 nm (ϵ, 81,000) and 440 nm (ϵ, 660). The infrared spectrum shows marked bands which can be assigned to *cis* and to *trans* double bonds. The n.m.r. spectrum of [16]annulene, at room temperature, is simple. It shows only one band, at τ3·27. This position is close to that observed for the olefinic protons of linear conjugated systems. When the temperature of the solution is lowered the band becomes progressively broader, and at $-110°$ the singlet is split into two main bands. One is a triplet at $\tau-0·43$, assigned to the four internal protons; the other is a multiplet at τ4·60, assigned to the external protons. This is a reversal of the behaviour shown by aromatic $(4n+2)$ systems.

Bisdehydro[16]annulene. Two bisdehydro[16]annulenes have been described in some detail. Isomer A has a typical non-aromatic ultraviolet absorption spectrum, with maxima at 283 and 297 nm. Its n.m.r. spectrum shows a quartet at τ2·25 assigned to four protons attached to two *trans* double bonds, and an octet at τ4·35 and a doublet at τ4·93. Isomer B also has a non-aromatic ultraviolet absorption spectrum, with maxima at 278 and 289 nm.

3.10. 18π-Electron systems.

[18]Annulene. Cyclo-octadecanonaene, or [18]annulene (CLXXVII) conforms to Hückel's rule ($n = 4$) and, provided the molecule can assume a planar configuration, should exhibit aromaticity. Molecular models and scale drawings indicate that the van der Waals radii of the six internal hydrogen atoms overlap slightly, but that this overlap is not so extreme as to exclude a reasonably planar carbon skeleton.

[18]Annulene has been synthesized, and has been found to display relatively little aromatic character in the classical sense. It is not particularly stable, for example, and it reacts readily with bromine and with maleic anhydride to give addition products. Various electrophilic substitution reactions were attempted, but these led to destruction of the hydrocarbon. On the other hand, the physical evidence clearly indicates that [18]annulene is aromatic. The empirical resonance energy has been determined from the heat of combustion, and found to be 100 ± 6 kcal mole^{-1}. X-ray analysis has established the fact that [18]annulene possesses a centre of symmetry, which rules out bond-alterna-

tion. The molecule has also been found to be essentially planar, with the cisoid bonds of bond length 1·419 (± 0·004) Å, and the transoid bonds of length 1·382 (± 0·003) Å. This is, of course, in marked contrast to cyclo-octatetraene which is not planar, and in which the carbon–carbon bonds alternate in length between 1·33 and 1·46 Å.

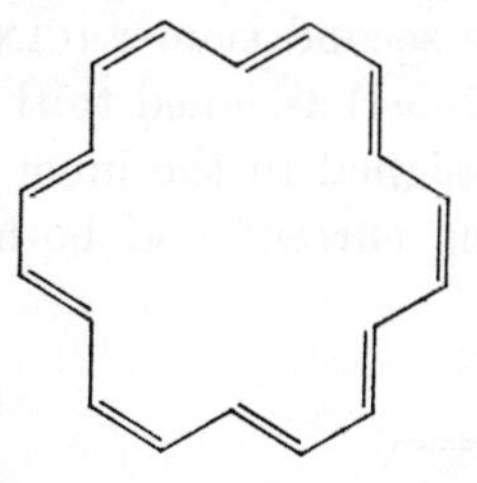

(CLXXVII)

The ultraviolet absorption spectrum of [18]annulene shows maxima at 278, 369, 408 and 448 nm. The n.m.r. spectrum shows two broad bands, one at very low field (τ1·1) due to the outer protons and one at very high field (τ11·8) due to the six internal protons. The areas under the two bands were found to be in the ratio 2:1. These peaks become progressively sharper and exhibit fine structure as the solution is cooled; in addition, the separation between the bands is increased. At −60° the low-field signal appears as a quartet-like multiplet centred at τ0·72, and the high-field signal as a complex multiplet centred at τ12·99. This increased separation of bands at low temperature is indicative of an increased ring current. On the other hand, heating a solution of [18]annulene causes the bands at τ1·1 and 11·8 to broaden, and at 40° these bands can no longer be recognized. At 110° the spectrum consists of a relatively sharp singlet at τ4·55. The ultraviolet spectrum was essentially unchanged over this temperature range, and it has therefore been concluded that [18]annulene possesses a ring current even in the temperature range in which the n.m.r. spectrum shows only a single peak. At higher temperatures it is thought that the protons change position at such a rate that an average value results for the band locations.

Several dehydro[18]annulenes are possible. In such compounds the shape of the molecule is affected by the presence of the formal triple bonds, but the 18 out-of-plane π-electron system is retained.

Two isomeric tridehydro[18]annulenes, for example, have been prepared; their ultraviolet absorption spectra are similar, and show considerable fine structure. In the n.m.r., the first isomer (CLXXVIII) gave a spectrum showing a complex band at low field (τ1·7–3·1) assigned to the outer protons, and a double doublet assigned to the inner protons at high field (τ8·26). The n.m.r. spectrum given by the second isomer (CLXXIX) similarly showed peaks at low field (τ1·7–3·4) assigned to the outer protons and at high field (τ7·8–8·6) assigned to the inner protons. Both isomers therefore sustain a ring current and both must be regarded as aromatic.

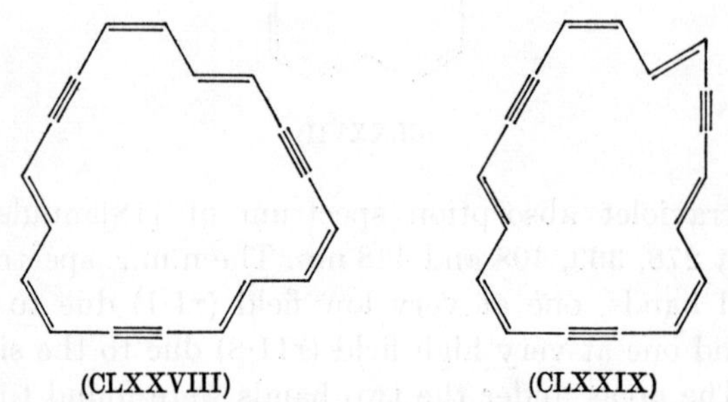

(CLXXVIII) (CLXXIX)

A tetradehydro[18]annulene (CLXXX) has also been described. The ultraviolet absorption spectrum of this brick-red compound is very similar to that given by tridehydro[18]annulene. The n.m.r. spectrum showed two multiplets, one at τ7·4–8·1 assigned to the inner (shielded) protons, and the other at τ1·8–3·5 assigned to the outer (deshielded) protons. It can be concluded therefore that this tetradehydro[18]annulene is aromatic.

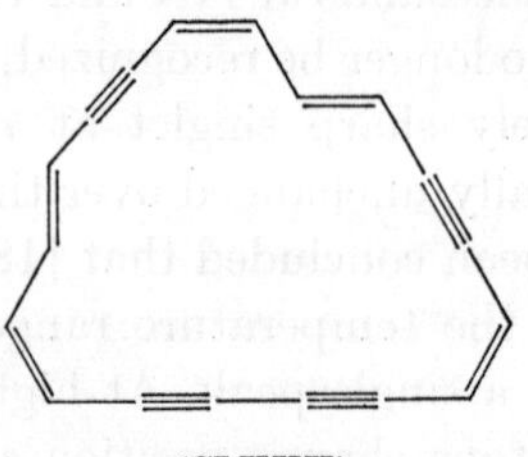

(CLXXX)

Hetero-substituted[18]annulenes. In [18]annulene there are 12 'outer' hydrogens and 6 'inner' hydrogens. Scale drawings suggest a small degree of overcrowding of the inner hydrogen atoms, but insufficient to exclude a reasonably planar carbon skeleton (Sondheimer, 1963); and the planarity of the molecule has been established experimentally.

Many hetero-substituted [18]annulenes are possible in which the 6 'inner' hydrogens are replaced by hetero-atoms. In [18]annulene trioxide (CLXXXI, Y = O), for example, all 6 'inner' hydrogens are replaced by three oxygen atoms; and similar molecules can be conceived in which the 6 'inner' hydrogens are replaced by three sulphur atoms, by three imino groups, or by any combination of oxygen atoms, sulphur atoms and imino groups.

Oxygen is a relatively small atom, and models and scale drawings suggest that the three oxygens in [18]annulene trioxide do not overlap, and therefore that this molecule should be planar. [18]Annulene trioxide has been synthesized (Badger, Elix & Lewis, 1966) and found to be a deep-red crystalline compound. The n.m.r. spectrum showed two peaks of equal area at very low field (τ1·32, 1·34), and it can be concluded that the molecule can sustain a significant induced ring current around the periphery. The ultraviolet absorption spectrum of [18]annulene trioxide supports the conclusion that this compound is aromatic. The spectrum is remarkably similar to that of the isoelectronic compound, tridehydro[18]annulene (fig. 3.1), and similar to that of [18]annulene itself. On the other hand [18]annulene trioxide reacted with bromine by addition and not by substitution.

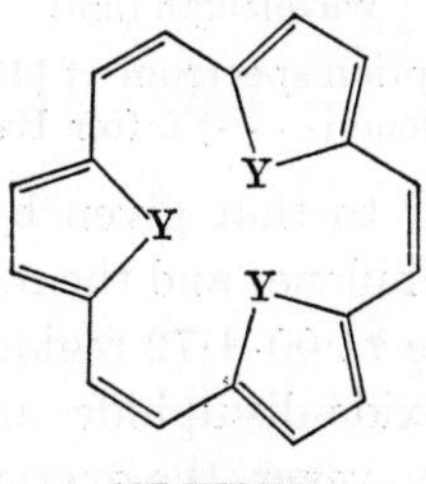

(CLXXXI)

Sulphur is a much larger atom than oxygen, and molecular models and scale drawings indicate that as each oxygen atom in [18]annulene trioxide is replaced by a sulphur atom there is a progressive increase in overcrowding. With [18]annulene dioxide-sulphide the overcrowding is not excessive, and this red compound strongly resembles [18]annulene trioxide. The ultraviolet absorption

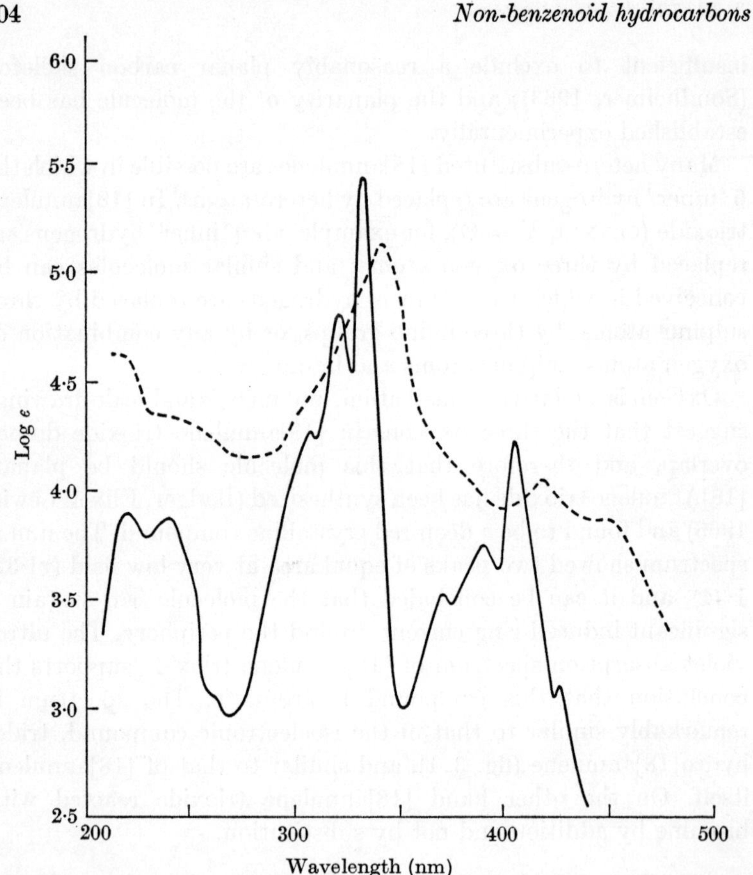

Fig. 3.1. Ultraviolet absorption spectrum of [18]annulene trioxide (——), and of tridehydro[18]annulene (- - - -) (after Badger, Elix & Lewis, 1966).

spectrum is very similar to that given by [18]annulene trioxide and by tridehydro[18]annulene; and the n.m.r. spectrum showed a series of peaks, all in the $\tau 1{\cdot}00$–$1{\cdot}72$ region.

With [18]annulene oxide-disulphide and [18]annulene trisulphide (CLXXXI, Y = S), however, the overcrowding is more serious. [18]Annulene oxide-disulphide is a yellow compound. The ultraviolet absorption spectrum (fig. 3.2) showed no fine structure; and the n.m.r. spectrum showed peaks at $\tau 2{\cdot}90$–$3{\cdot}39$, that is, at much higher field than for [18]annulene trioxide. Similarly, [18]annulene trisulphide is a yellow, stable, crystalline compound, and its ultraviolet absorption spectrum is simple and at slightly shorter wavelength than that of the oxide-disulphide (fig. 3.2). The

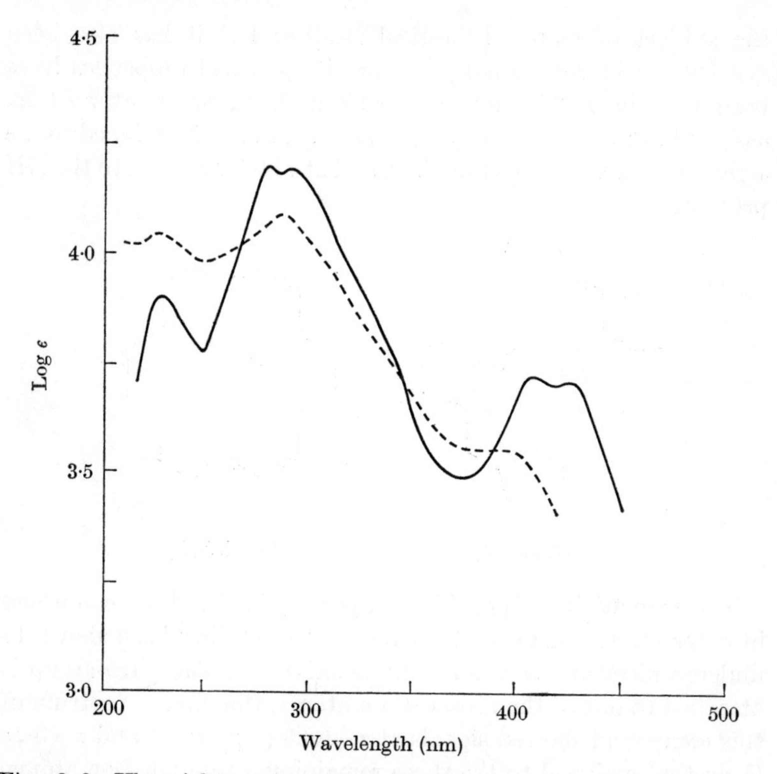

Fig. 3.2. Ultraviolet and visible absorption spectra of [18]annulene 1,4-oxide-7,10:13,16-disulphide (——), and of [18]annulene 1,4:7,10:13,16-trisulphide (- - - -), both in 95% ethanol (after Badger, Lewis & Singh, 1966).

n.m.r. spectrum shows two signals of equal areas at τ3·27 and 3·33. It seems therefore that [18]annulene oxide-disulphide and [18]annulene trisulphide are non-planar, non-aromatic systems in which the three heterocyclic rings are linked by three essentially olefinic vinylene groups. The trisulphide has been shown experimentally to be non-planar: the brucine salt of triepithio[18]annulene-5,11,18-tricarboxylic acid exhibits marked and rapid mutarotation in solution. Finally, [18]annulene trisulphide reacted with bromine by addition.

Porphin and porphyrins. Porphin (CLXXXII), the parent ring-system of the porphyrins, can be regarded as a derivative of the hypothetical 1,10-diaza[18]annulene (CLXXXIII). Porphin has been the subject of many theoretical studies; but it has also been synthesized in very small yield and its physical properties have been examined. The n.m.r. spectrum shows bands at $\tau -1{\cdot}22$, assigned to the four *meso* hydrogen atoms, at $\tau 0{\cdot}08$, assigned to the eight protons on the pyrrole rings and at $\tau 14{\cdot}4$, assigned to the NH protons.

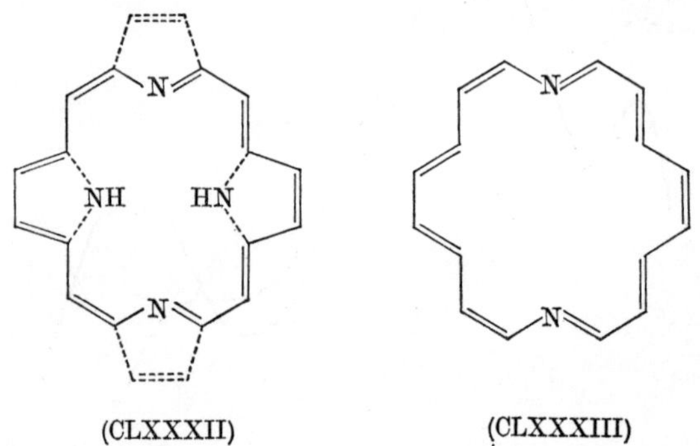

(CLXXXII) (CLXXXIII)

The aromaticity of porphin and porphyrins has been confirmed in other ways. For example, octaethylporphin has been shown to undergo nitration to a mono-nitro-derivative. The nitro-group is attached to one of the *meso* carbon atoms; the n.m.r. spectrum of this compound showed signals at $\tau -0{\cdot}85$ (2 protons) and $\tau -0{\cdot}76$ (1 proton) assigned to the three remaining *meso*-hydrogen atoms.

The X-ray crystal structure of porphin has been determined, and the molecule found to be nearly planar. Surprisingly, however, the average length of the C_β—C_β bonds was found to be 1·342 Å, which is very close to that for a double bond.

Many X-ray structural investigations of porphyrins have been undertaken. Among the metal-complexed tetraphenylporphins, for example, it seems that the molecular configuration changes from a planar to a buckled non-planar one depending on the crystal packing, and on the substituents attached to the porphin nucleus. Copper and palladium tetraphenylporphin are non-planar, but the porphin nucleus of the zinc compound is planar. Iron porphin was found to be nearly planar except that the iron atom was 0·2 Å above the plane of the four nitrogen atoms. Crystalline tetraphenylporphin is also non-planar. It seems that the porphin skeleton can exist in a planar or buckled form depending on the environment.

Dihydroporphins (chlorins and rhodins) and tetrahydroporphins (bacteriochlorins) may also be regarded as derivatives of 1,10-diazo[18]annulene (CLXXXIII); indeed they are more closely related than porphin itself. It is of interest that chlorins are extraordinarily susceptible to electrophilic attack at the γ and δ bridge positions.

3.11. 20π-Electron systems.

[20]Annulene. This compound does not conform to Hückel's rule. It is a very unstable hydrocarbon, being readily decomposed by air or sunlight. Its ultraviolet absorption spectrum showed high-intensity peaks at 267, 284 and 297 nm, and low-intensity maxima at 375 and 396 nm. Bisdehydro[20]annulene has also been described.

3.12. 24π-Electron systems.

[24]Annulene. This compound does not conform to Hückel's rule. It is very unstable, being readily decomposed by air and sunlight. The n.m.r. spectrum shows a single band at τ3.16 confirming the absence of aromaticity. At low temperatures, however, the inner protons give rise to bands at $\tau - 2{\cdot}8$ to $-1{\cdot}2$, and the outer protons to a band at $\tau 5{\cdot}27$. This behaviour is similar to that observed with [16]annulene and is a reversal of that observed with aromatic $(4n+2)$ systems. The ultraviolet absorption spectrum of [24]annulene is relatively simple, with maxima at 264, 350, 363 and 512 nm.

Tetradehydro[24]annulene has also been obtained. The n.m.r. spectrum shows a double doublet at $\tau 1{\cdot}80$ assigned to the inner protons, and a complex band at $\tau 3{\cdot}6$–$4{\cdot}8$ assigned to the outer protons. The absence of absorption at high field is in accord with the absence of an induced ring current.

3.13. 30π-Electron systems.

[30]Annulene. This compound conforms to Hückel's rule ($n = 7$). It has been synthesized, but it is unstable, and its ultraviolet absorption spectrum is simple, with maxima at 329 and 431 nm. It seems that this ring system follows the Longuet-Higgins & Salem rule (1959, 1960), namely that, as the ring becomes very large, the $(4n+2)$ rule no longer operates. [30]Annulene appears to be a cyclic polyene.

3.14. Examination of Hückel's rule.

Hückel's rule predicts that conjugated monocyclic systems which contain $(4n+2)$ π-electrons will be aromatic, and that systems which contain $4n$ π-electrons will be non-aromatic. This rule was enunciated in the nineteen thirties, but many years elapsed before it could be put to the test of experiment. Several new polymethine systems have now been prepared and a detailed examination of the rule can be carried out. The cyclopropenium cation system has been shown to be aromatic, and cyclobutadiene to be non-aromatic. The cyclopentadienide anion, benzene, and the cyclo-heptatrienium cation are aromatic; but cyclo-octatetraene is non-aromatic. Derivatives of [10]annulene which have a planar perimeter are aromatic, but [12]annulene and tridehydro[12]annulene are not aromatic. Planar [14]annulenes, such as bisdehydro[14]annulene, are aromatic; but [16]annulene is not aromatic. [18]Annulene and its planar derivatives are aromatic; but [20]annulene and [24]annulene are not aromatic; and [22]annulene and [26]annulene are unknown. All these findings are in accord with Hückel's rule. Only [30]annulene seems to be at variance. This hydrocarbon conforms to Hückel's rule ($n = 7$), and aromaticity is therefore predicted, but its properties suggest that it is a cyclic polyene.

With this exception the agreement is surprisingly good. It has been pointed out (Chung & Dewar, 1965) that calculations by the simple Hückel method indicated that systems with $4n$ π-electrons should have lower resonance energies per atom than $(4n+2)$ systems, but they did not imply that the former should be as unfavourable as experiment has now established.

The problem has recently been re-examined (Dewar & Gleicher, 1965*a*) using greatly refined methods and eliminating some simplifying assumptions. π-Binding energies ($E_{\pi b}$) and resonance energies (E_R) have been calculated for the annulenes having certain specified geometries. The results are given in table 3.1, and have been plotted as a function of ring size in fig. 3.3.

The alternation in resonance energy with ring size is striking. All the annulenes having $4n$ π-electrons are found to have negative resonance energies and must therefore be non-aromatic. All the annulenes having $(4n+2)$ π-electrons, up to and including [22]annulene (which is still unknown), are predicted to be aromatic. The

TABLE 3.1. *π-Binding energies and resonance energies of the annulenes*

(Dewar and Gleicher, 1965*a*)

Annulene	$E_{\pi b}$, e.v.	E_R, e.v.
Cyclobutadiene	3·091	−0·815
Benzene	7·177	1·318
Cyclo-octatetraene	7·485	−0·327
[10]Annulene	10·844	1·079
[12]Annulene	11·667	−0·051
[14]Annulene	14·294	0·623
[16]Annulene	15·489	−0·135
[18]Annulene	17·854	0·277
[20]Annulene	19·215	−0·315
[22]Annulene	21·573	0·090
[24]Annulene	22·901	−0·535
[26]Annulene	25·160	−0·229
[28]Annulene	26·977	−0·365
[30]Annulene	28·824	−0·471

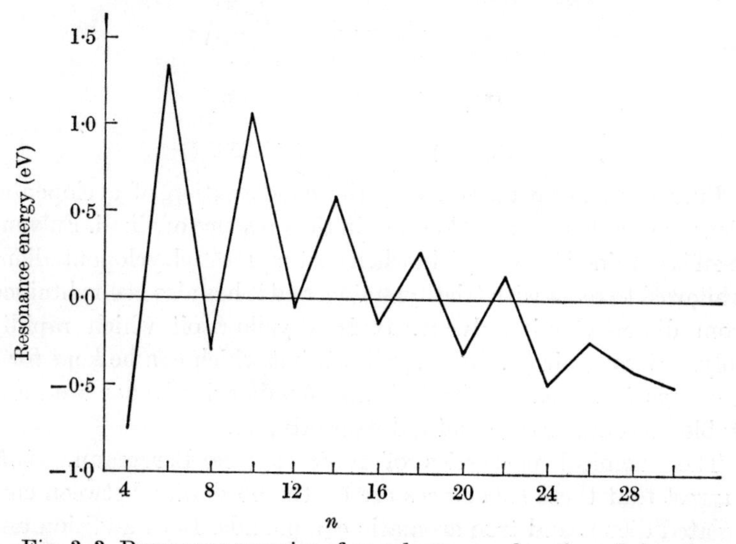

Fig. 3.3. Resonance energies of annulenes as a function of ring size (after Dewar & Gleicher, 1965*a*).

higher annulenes having $(4n+2)$ π-electrons, that is [26]annulene and [30]annulene, are predicted to be non-aromatic and to disobey Hückel's original rule. The graphical representation provides a clear explanation for the cut-off in aromaticity in $(4n+2)$ compounds first suggested by Longuet-Higgins & Salem (1959, 1960).

4. More complex systems

4.1. Fulvenes and fulvalenes. The non-alternant hydrocarbon fulvene (CLXXXIV) may be considered as a derivative of the cyclopentadienide anion, by virtue of the contribution of (CLXXXIV*b*) to the resonance hybrid. Measurements of the dipole moments of fulvene, and of substituted fulvenes, have confirmed the implied separation of charge. Fulvene itself (CLXXXIV) has a dipole moment of 1·1*D*, dialkylfulvenes of *ca.* 1·5*D*, and 6,6-diphenylfulvene of 1·9*D*.

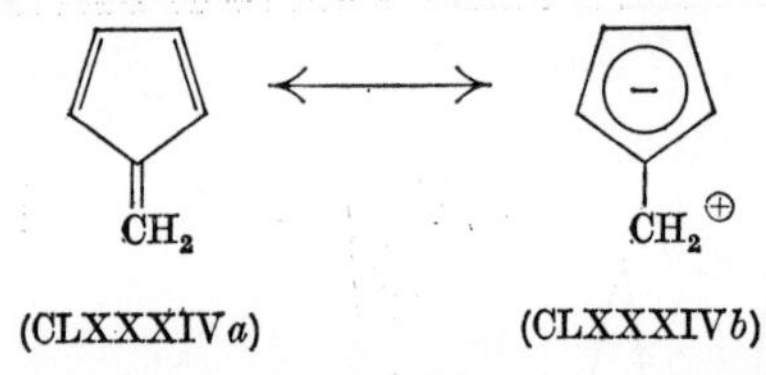

(CLXXXIV*a*) (CLXXXIV*b*)

Fulvenes can be prepared by the condensation of cyclopentadiene with aldehydes and ketones in the presence of alkali. Fulvene itself is obtained in good yield when acetoxymethylcyclopentadiene is allowed to react with triethylamine, and it has also been obtained from dimethylaminofulvene. It is a yellow oil which rapidly polymerizes in air (L. *fulvus*, yellow), but which can be kept for a few days *in vacuo*. 6,6-Dialkyl- and 6,6-diaryl-fulvenes are more stable (and are orange and red respectively).

The chemical properties of fulvenes (see Bergmann, 1955) suggest that these substances are on the borderline between conjugated dienes and true aromatic compounds. Both addition and substitution reactions occur; and fulvenes readily undergo the Diels–Alder reaction. The physical evidence confirms the small degree of aromaticity in fulvenes. The ultraviolet absorption spectrum of fulvene shows maxima at ~241 and ~360 nm; and fulvenes generally exhibit strong absorption at about 270 nm with a weaker band at about 365 nm, with tailing into the visible region. The n.m.r. spectrum of fulvene indicates that any ring current must be so small as to be almost negligible. With fulvene itself the ring protons resonate at $\tau 3{\cdot}56$ and $\tau 3{\cdot}89$, and the exocyclic protons at $\tau 4{\cdot}22$. Substituted fulvenes give similar spectra. The spectrum of 6,6-diphenylfulvene has been shown to be a perfectly symmetrical AA′BB′ type with the fulvene ring protons resonating at $\tau_A 3{\cdot}61$ and $\tau_B 3{\cdot}895$. Similarly, 6-phenylfulvene (in acetone) gives ring proton resonances at $\tau_A 3{\cdot}69$, $\tau_{A'} 3{\cdot}29$, $\tau_B 3{\cdot}54$ and $\tau_{B'} 3{\cdot}37$. These values may be compared with that given for cyclopentadiene ($\tau 3{\cdot}58$).

It is, perhaps, more instructive to compare the *ortho* coupling constants. Various cyclohexenes show olefinic proton couplings of 9–11 cps, and substituted benzenes exhibit *ortho* coupling constants in the range 5–9 cps; the olefinic proton coupling in cyclopentene is 5·1–5·4 cps; the *ortho* coupling constant in the five-membered ring in azulene is 3·5 cps and in the pentalenide dianion is 3·0 cps. In the fulvenes, however, the *ortho* coupling constant for the 1,2 bonds is *ca.* 5·5 cps, and for the 2,3 bonds is near 2·0 cps. For 6-phenylfulvene, for example, the coupling constants are $J_{1,2}$ 5·57 and $J_{2,3}$ 1·88. The latter figure may be compared with the corresponding coupling constant ($J_{3,4}$ 3·5 cps) for heteroaromatic compounds such as furan, pyrrole and thiophen, and with the corresponding coupling constant ($J_{3,4}$ 1·94 cps) for cyclopentadiene. It can be concluded from these data that fulvenes possess little aromaticity. Indeed, it may be noted that the 2,3 coupling constant (1·92 cps) for 6-dimethylaminofulvene (CLXXXV) is only slightly greater than the corresponding constant (1·88 cps) for 6-phenylfulvene. This is of some interest in view of the increased participation of the ionic structure (CLXXXV*b*) in the resonance hybrid. This increased participation is confirmed by the fact that 6-dimethylaminofulvene has a dipole moment of 4·5*D*.

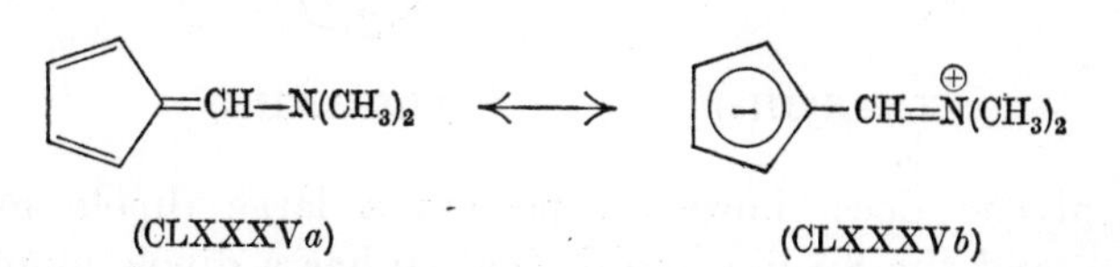

Calculation of the delocalization energy of fulvene gives a value of $1{\cdot}466\beta$, intermediate between that for butadiene ($0{\cdot}472\beta$) and for benzene ($2{\cdot}00\beta$). However the strain energy for fulvene is considerable, and must be deducted from the delocalization energy to give a figure (13 kcal mole^{-1}) for the resonance energy. Experimentally the empirical resonance energy of fulvene has been determined from heats of combustion data to be about 12 kcal mole^{-1}.

Fulvalene (CLXXXVI) was first discussed as a possible aromatic system in 1950 (Brown) and many attempts have been made to synthesize it. Dilute solutions of fulvalene have been prepared, but the hydrocarbon is too unstable to be isolated. On the other hand, octachlorofulvalene (CLXXXVII) has been prepared; it is a yellow unreactive solid which does not take part in the Diels–Alder synthesis. Similarly, a hexaphenylfulvalene has been prepared. This is a dark khaki solid, and its ultraviolet and visible absorption spectrum shows maxima at 264 and 412 nm.

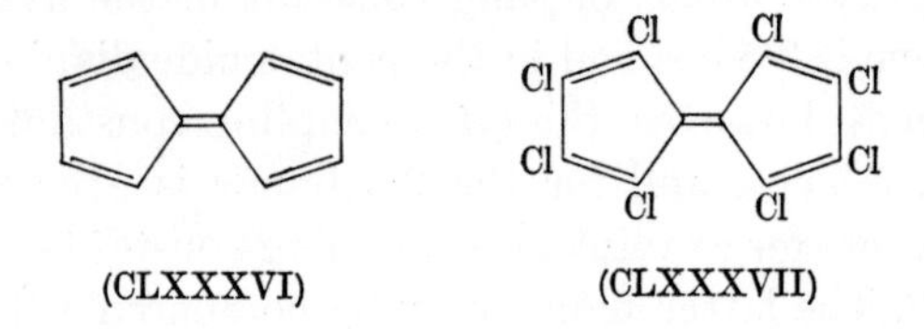

(CLXXXVI) (CLXXXVII)

Compounds analogous to the fulvenes and fulvalenes, but having different-sized rings are also possible. Heptafulvene (CLXXXVIII), for example, may be considered as a derivative of the tropylium cation. It has been synthesized, but determination of the resonance energy from the heat of hydrogenation showed that the molecule does not possess a high degree of π-electron delocalization. The ultraviolet absorption spectrum shows intense absorption in the 240–300 nm region, and other maxima at 500, 545 and 600 nm.

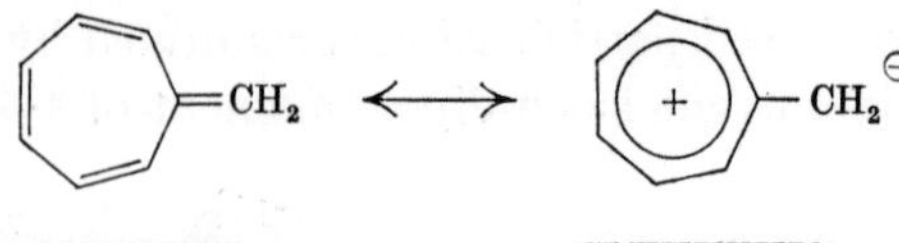

(CLXXXVIII*a*) (CLXXXVIII*b*)

Heptafulvene does, however, possess a large dipole moment (3·07*D*), and the 8,8-dicyano derivative has a dipole moment of 7·49*D*. 8-Vinylheptafulvene and heptafulvalene (CLXXXIX) have also been prepared, and their n.m.r. spectra are consistent with the view that neither possesses substantial π-electron delocalization. Similarly, 1,2-benzoheptafulvene and 3,4-benzoheptafulvene have been prepared, but again the n.m.r. spectra suggest that dipolar resonance interactions are of minor importance in the ground state.

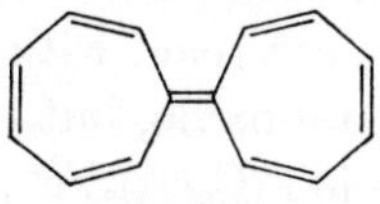

(CLXXXIX)

4.2. Metallocenes. When phenylmagnesium bromide is treated with ferric chloride, biphenyl is obtained in almost quantitative yield. It seemed possible that similar treatment of cyclopentadienyl magnesium halide might give fulvalene; but a crystalline organo-iron compound, $(C_6H_5)_2Fe$, was obtained instead (Kealy & Pauson, 1951). The same compound was obtained soon afterwards by the actions of cyclopentadiene on reduced iron at 300° (Miller, Tebboth & Tremaine, 1952).

This organo-iron compound attracted considerable interest. A sandwich-type structure (CXC) was soon suggested; and the substance was recognized as a new aromatic system and named ferrocene shortly afterwards. Since then a new branch of aromatic chemistry has been opened up (see Fischer, 1955; Pauson, 1955*b*, 1959).

The sandwich-type structure (CXC) has been confirmed by X-ray crystallographic analysis. In the crystal, the two rings are arranged so that an anti-prism is formed; but it seems that, in solution, the two rings rotate quite freely. It may be noted, however, that many bridged ferrocenes have been prepared.

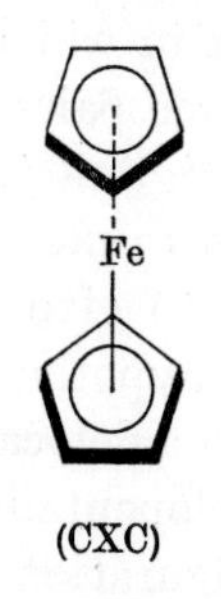

(CXC)

The properties of ferrocene are certainly consistent with formulation as an aromatic compound. It forms orange needles, m.p. 173–174°, b.p. 249°, and is soluble in all common solvents. It is volatile above 100° at atmospheric pressure, and is stable to at least 450° in the vapour phase. It can be distilled in steam, and is not attacked by 10 % sodium hydroxide, or by boiling concentrated hydrochloric acid. It does not react with maleic anhydride, is not hydrogenated over platinum oxide, but is readily oxidized in acid solution to the less stable ion $(C_5H_5)_2Fe^+$, which can be isolated as a sparingly soluble picrate. Ferrocene undergoes alkylation, acylation, sulphonation, metalation, arylation, formylation, aminomethylation and other reactions characteristic of a highly reactive aromatic system. Competitive experiments, using the Friedel–Crafts acylation reaction, lead to the following order of reactivity:

$$\text{ferrocene} > \text{anisole} > \text{benzene}.$$

Ferrocene is diamagnetic and has zero dipole moment. In the infrared it gives a band at 3075 cm^{-1}, assigned to C—H stretching, and this is the region typical for aromatic C—H bonds.

In the n.m.r. spectrum the protons of ferrocene appear at $\tau 6{\cdot}01$; that is, at relatively high field, and suggesting a negatively charged ring. Similarly, ruthenocene gives an n.m.r. spectrum with proton signals at $\tau 5{\cdot}61$; and magnesocene with signals at $\tau 4{\cdot}30$.

Finally, it may be noted that the empirical resonance energy of ferrocene has been determined, from the heat of formation, to be 113 kcal $mole^{-1}$.

The undoubted aromaticity of ferrocene leads to the problem of the electronic structure, and it must be admitted that the precise nature of the metal-ring bond is still in dispute. According to Fischer and his co-workers, ferrocene should be regarded as a new type of penetration complex in which two staggered cyclopentadienide anions (each with three pairs of π-electrons) are separated by a central Fe^{2+} ion which interacts with each pair forming six coordinate covalencies. This results in a complete filling of the hybrid d^2sp^3 metal orbitals to give the inert gas configuration. According to the alternative hypothesis, the ring-metal bonding in ferrocene results largely from a delocalized covalent bond between the metal atom and two cyclopentadienide radicals.

The outer electronic configuration of Fe in the ground state is $3d^64s^2$. If a $4s$ orbital and a $3d$ orbital are hybridized, two non-equivalent ds orbitals are obtained. One of these, $ds_{(1)}$, is of slightly lower energy than a $3d$ orbital; the other, $ds_{(2)}$, is of higher energy, approximately that of a $4p$ orbital. In ferrocene, the Fe atom is assumed to have the configuration $ds^2_{(1)}\ 3d^6$ (see below), with two unpaired electrons in $3d$ orbitals. With this configuration, bonding can occur by the overlap of a singly-filled molecular orbital of a cyclopentadienyl radical with a singly-filled $3d$ orbital of the iron atom. Thus ferrocene would have no unpaired electrons and it is, of course, diamagnetic.

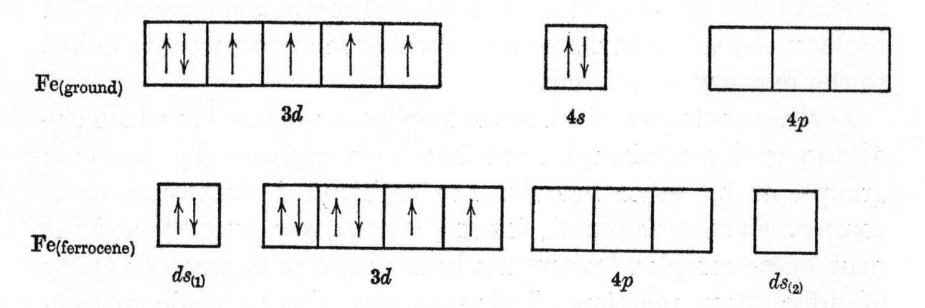

Whether ferrocene is to be regarded as being derived from two cyclopentadienide anions and Fe^{2+}, or from two cyclopentadienyl radicals and Fe, is immaterial for the present discussion. The two hypotheses differ mainly in the degree to which the π-electrons are thought to participate in the bonding with metal orbitals.

Many analogues of ferrocene have been prepared. Nearly all the elements of the three transition series have been shown to form di-π-cyclopentadienyl compounds of general formula

$$[(C_5H_5)_2M^n]X_{n-2}$$

where n is the oxidation number of metal M, and X is a univalent anion. In the first transition series, compounds of type $(C_5H_5)_2M$ have been described for all metals from titanium to nickel. Most of these have the same melting point as ferrocene, and form a series of isomorphous crystals. With the exception of that derived from manganese, these compounds must be regarded as similar to ferrocene in electronic structure. The uniqueness of the manganese compound has been attributed to the peculiar stability of the electronic configuration of the manganous ion: biscyclopentadienylmanganese has an ionic structure, $(C_5H_5^-)_2Mn^{2+}$. Titanium and vanadium both form two isomeric cyclopentadienyl derivatives, one of which in each case is an ionic compound and the other a compound having ferrocene-like ring-metal bonding.

Many elements outside the transition series also form cyclopentadienyl derivatives, but not of the ferrocene type. The alkali metals form cyclopentadienides $(C_5H_5^-)M^+$; the alkaline earths form $(C_5H_5^-)_2M^{2+}$; and the rare earths, $(C_5H_5^-)_3M^{3+}$. Some of these compounds, nevertheless, show some similarities to ferrocene: biscyclopentadienylmagnesium, for example, has an ionic structure but it has, in the solid state, the same sandwich structure as ferrocene. A third type of biscyclopentadienyl compound is exemplified by biscyclopentadienylmercury in which the two rings are linked to the mercury by σ-bonds.

Further analogues of ferrocene have been prepared in which one of the cyclopentadienyl rings has been replaced by carbonyl groups, or by other ligands. The carbonyl derivatives include compounds such as (CXCI), (CXCII), (CXCIII) and (CXCIV); and the manganese complex (CXCIV) has been shown to undergo a variety of substitution reactions. Reference may also be made to compounds such as (CXCV) and (CXCVI).

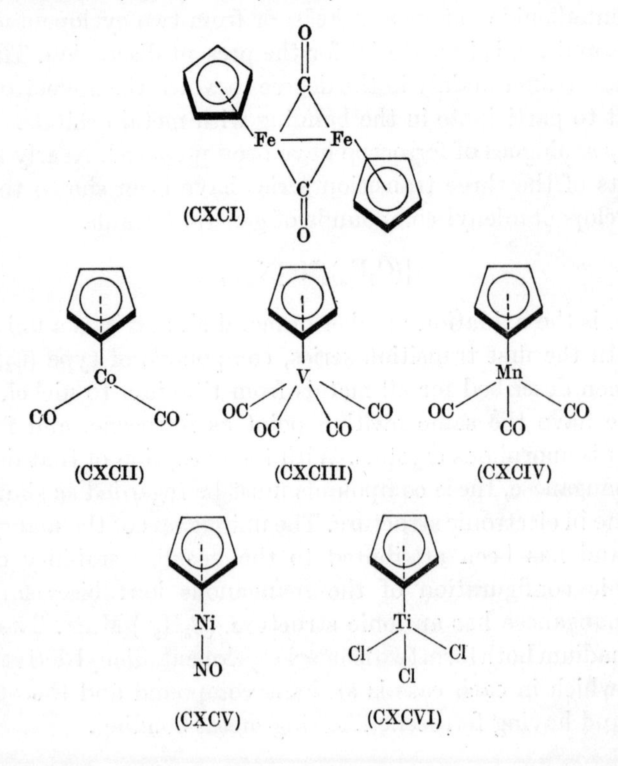

Further, the cyclopentadienide anion has a sextet of π-electrons, and it might therefore be expected that other ring systems having a sextet of π-electrons would form sandwich compounds with transition metals. Specifically, it might be expected that the cyclobutadienide dianion, benzene, the tropylium cation, and the cyclo-octatetraenium dication might form ferrocene-like compounds. Furthermore, ferrocene-like compounds might be expected from heterocyclic ring systems, such as thiophen, and some of these have been shown to exist.

Examples of these compounds are the water-soluble bis-benzenechromium cation (CXCVII) and the brown-black bis-benzenechromium (CXCVIII) which is sparingly soluble in most organic solvents. X-ray and electron diffraction investigations of bis-benzenechromium have confirmed that there is no alternation in the lengths of the carbon–carbon bonds. Moreover, no bond alternation was found in 9,10-dihydrophenanthrenechromium tricarbonyl or in phenanthrenechromium tricarbonyl. Tricarbonylbenzenechromium (CXCIX), tricarbonylthiophenchromium (CC) and tricarbonyltropyliummolybdenum (CCI) have been obtained, and the X-ray crystal structure of the first of these has been determined. Chromium and tungsten derivatives of the tropylium cation have also been prepared. The structure of the tricarbonyltropylium iron cation seems, however, to be such that the iron is bonded to five carbon atoms leaving one double bond not involved in coordination to the metal.

Special interest attaches to the paramagnetic cyclopentadienylcycloheptatrienylchromium cation (CCII) and the diamagnetic cyclopentadienylcycloheptatrienylchromium (CCIII). The latter is isoelectronic with dibenzenechromium. The n.m.r. spectrum showed only two types of proton in the ratio 5:7, with τ values of 6·90 and 5·08 respectively. The dipole moment is $0{\cdot}79 \pm 0{\cdot}05$ in benzene, and $0{\cdot}73 \pm 0{\cdot}05D$ in cyclohexane, proving that there is a slight charge separation in the complex. This may be interpreted in the sense of structure (CCIII).

Another compound of special interest is cyclopentadienylhexakistrifluoromethylbenzenerhodium: in this compound the benzene nucleus is non-planar. Only four atoms of the benzene nucleus are involved in bonding to the metal ion; C_1 and C_4 form σ bonds, and a π bond from C_2—C_3 completes the metal

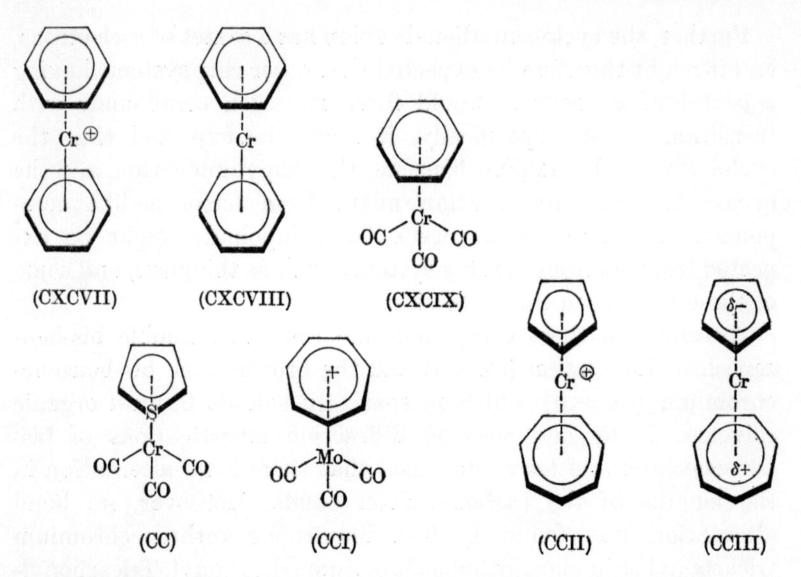

co-ordination. The benzene ring is 'hinged' across C_1 and C_4; and the non-co-ordinated double bond (C_5—C_6) is $1{\cdot}32 \pm 0{\cdot}05$ Å.

4.3. Bicyclic systems.

Pentalene. Pentalene (CCIV) was first postulated as a possible aromatic system in 1922 (Armit & Robinson), but according to Craig's rule should be non-aromatic. There have been many attempts to prepare pentalene, and although some pentalene derivatives have been obtained it seems that pentalene itself cannot be particularly stable. Moreover, the diketone (CCV) has been shown to exist as such and not as a dihydroxypentalene; the pentalene structure cannot therefore have any significant resonance energy.

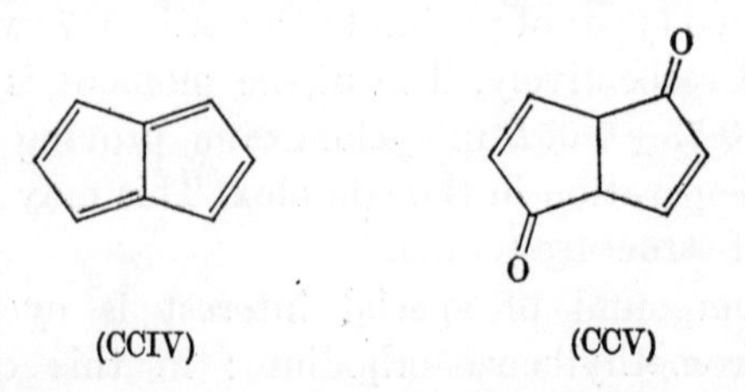

1,2-4,5-Dibenzopentalene (CCVI) has been prepared, and derivatives of dibenzopentalene have been known for many years. Dibenzopentalene is a bronze hydrocarbon which behaves like a conjugated diene. It reacts with bromine by addition, is readily reduced with sodium and ethanol, and is very readily ozonized.

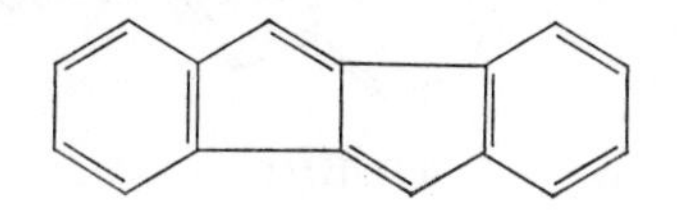

(CCVI)

Hexaphenylpentalene has also been synthesized. It was obtained as green-brown needles, stable in the solid state, but its solutions were found to be sensitive to air. Several heterocyclic analogues of pentalene can also be conceived. Indolo[3,2-*b*]indole (CCVII) is of special interest and it has been synthesized.

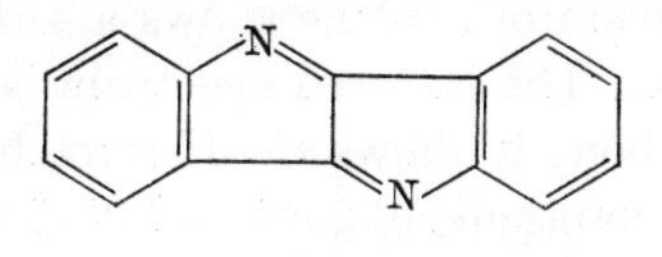

(CCVII)

Although pentalene is not aromatic the pentalenide dianion, having 10 π-electrons, might be expected to display aromaticity. This dianion has been prepared, and its n.m.r. spectrum shows a triplet at $\tau 4{\cdot}27$ and a doublet at $\tau 5{\cdot}02$, with identical splittings of 3·0 cps. These shifts are in approximate accord with expectation based on a ring current deshielding effect of π-electrons, and a shielding effect associated with the negative charge on the carbon atoms. The ultraviolet absorption spectrum of the dianion shows a maximum at 296 nm and a slight shoulder at 210 nm. Theoretical studies of the pentalenide dianion have been recorded.

Azulene. Azulenes have been known for a very long time. The blue colour of camomile oil was first reported in the fifteenth century, and it has since been found that about 20 % of the essential oils which have been investigated contain small quantities of a blue substance. The isolation of these blue substances, the azulenes, was made possible by the fact that they are soluble in concentrated acids, from which solutions they separate on dilution with water. The structure of the parent hydrocarbon (CCVIII) was elucidated in 1936 (St Pfau & Plattner); and the fact that many substituted azulenes occur in nature, combined with the interest of these compounds to aromatic chemistry, has resulted in an extensive literature.

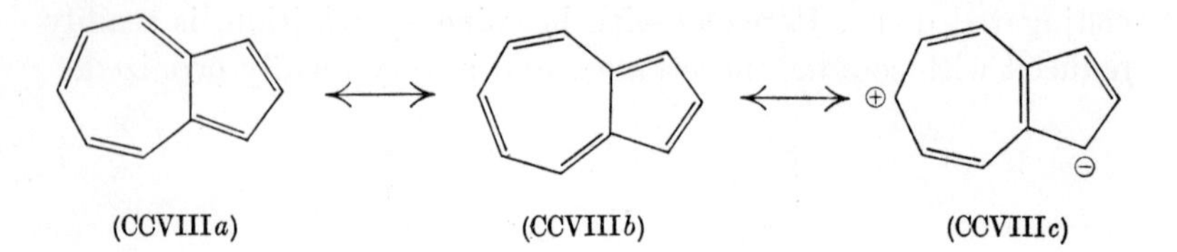
(CCVIII*a*) (CCVIII*b*) (CCVIII*c*)

There is no doubt that azulene must be considered an aromatic compound. It undergoes a wide variety of electrophilic substitution reactions to give 1-substituted azulenes, and 1,3-disubstituted azulenes undergo substitution at the 5-position. Detailed studies of the protonation of azulenes have shown that the process is reversible, and occurs at the 1-position.

The n.m.r. spectrum of azulene shows a series of bands in the τ1·84 to 3·06 region. The infrared spectrum is also typical of an aromatic hydrocarbon; it shows C—H stretching bands at 3030 and 3086 cm^{-1} (*cf.* napthalene, 2994 and 3067 cm^{-1}).

The ultraviolet and visible absorption spectrum of azulene shows a considerable amount of fine structure, particularly at $-170°$ (fig. 4.1). In the ultraviolet region the introduction of a methyl substituent results in a bathochromic shift, which is larger for substituents on the five-membered ring than for substituents on the seven-membered ring, but not all the bands are affected to the same extent. In the visible region of the spectrum, however, alkylation in the 1, 3, 5 and 7 positions causes a bathochromic shift, and alkylation in the 2, 4, 6 and 8 positions a hypsochromic shift. Azulenes in the former group therefore tend to be blue, and azulenes in the second group are violet.

The resonance energy of azulene has been determined, and its magnitude serves to confirm the conclusion that this non-alternant hydrocarbon is aromatic. Azulene itself has an empirical resonance energy of 46 kcal $mole^{-1}$. This value is in excellent agreement with the value (45·5 kcal $mole^{-1}$) determined by comparison of the heats of combustion of the two isomeric compounds, cadalene (CCIX) and guaiazulene (CCX).

The crystal structure of azulene is disordered and it has not been possible to determine the bond lengths with complete accuracy. However the transannular carbon–carbon bond (1·482 Å) is distinctly longer than the other carbon–carbon bonds (1·383–1·413 Å), all of which are in the usual aromatic range. Finally, the

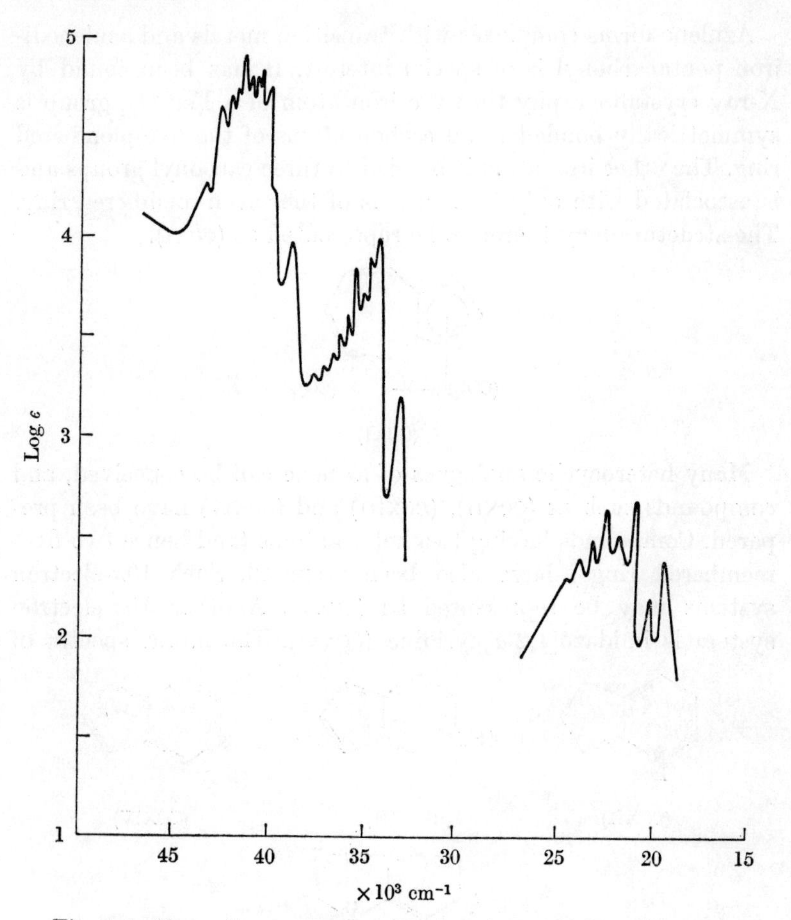

Fig. 4.1. Absorption spectrum of azulene in methanol–ethanol at −170° (after Clar, 1950).

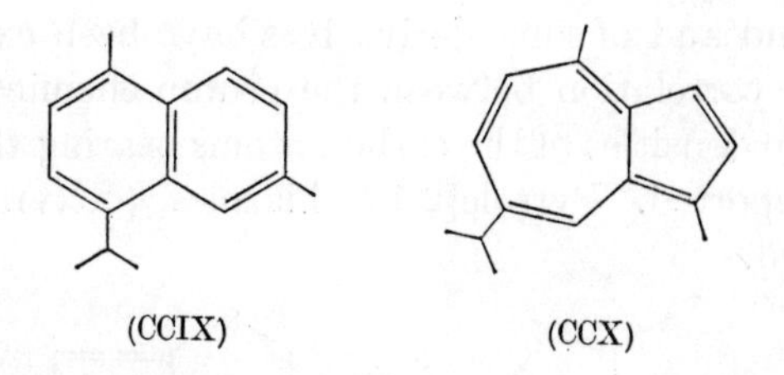

dipole moment of azulene has been found to be 1·0*D* confirming contributions from resonance structures having a separation of charge (CCVIII*c*).

Azulene forms complexes with transition metals and azulenediiron pentacarbonyl is of special interest. It has been found by X-ray crystallography that the iron atom of a $Fe(CO)_2$ group is symmetrically bonded to all carbon atoms of the five-membered ring. The other iron atom is bonded to three carbonyl groups and is associated with only three atoms of the seven-membered ring. The structure may therefore be represented as (CCXI).

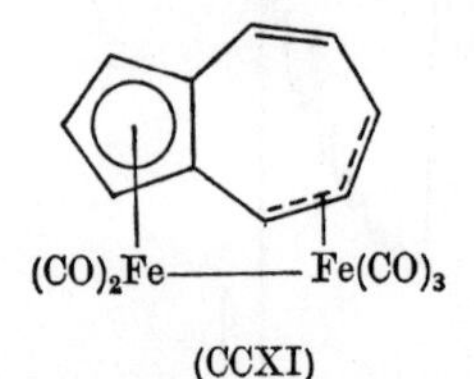

(CCXI)

Many heterocyclic analogues of azulene can be conceived, and compounds such as (CCXII), (CCXIII) and (CCXIV) have been prepared. Compounds having two hetero-atoms (and hence two five-membered rings) have also been prepared. Such 10π-electron systems may be represented as (CCXV). Another 10π-electron system is imidazo[1,2-*a*]pyridine (CCXVI). The n.m.r. spectra of

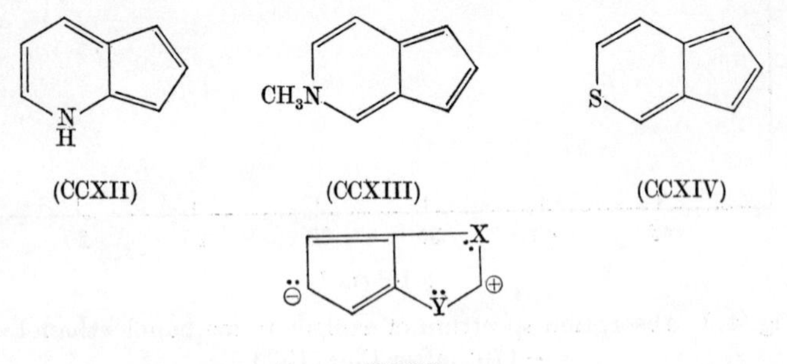

(CCXII) (CCXIII) (CCXIV)

(CCXV)

this compound and of some derivatives have been examined and a qualitative correlation between the proton chemical shifts and the π-electron densities of the carbon atoms bearing three protons has been reported. Pyrrole[2,1-*b*]thiazoles (CCXVII) have also been prepared.

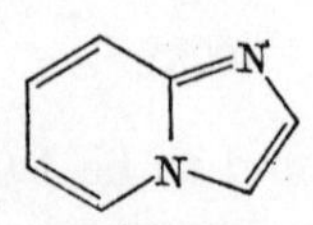

(CCXVI)

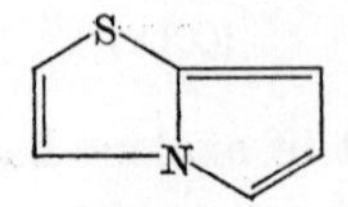

(CCXVII)

Vinologue of tropone. Tropone is a derivative of the cyclohepta-trienium cation and is an aromatic system. An analogous system (CCXVIII) may be regarded as a vinologue of tropone. The parent compound has been synthesized, but seems to be stable only in solution. A methoxy derivative has also been prepared, and found

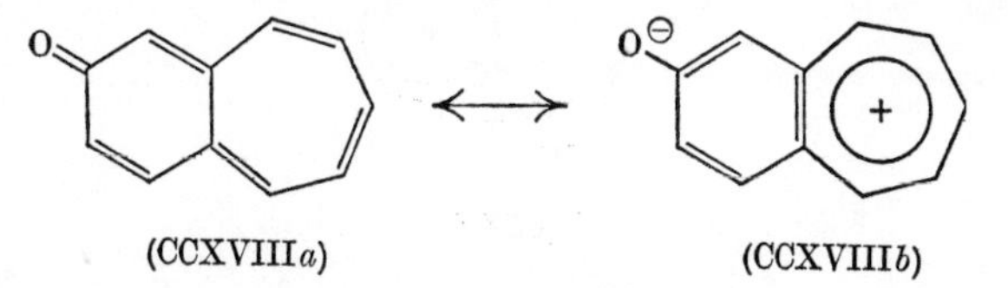

(CCXVIII*a*) (CCXVIII*b*)

to be much more stable. The ultraviolet and visible absorption spectrum showed maxima at 220, 267, 302, 347, 418 (infl), 434 and 494 nm (log ϵ respectively 4·1, 4·4, 4·3, 4·2, 3·6, 3·6 and 3·6). The n.m.r. spectrum (in $CDCl_3$) showed a signal at τ6·0 assigned to the methoxyl protons, and a band at τ2·0 to 3·1 assigned to the ring protons. It is reasonable to conclude that the whole nucleus can sustain a ring current, and the system may be regarded as aromatic.

Heptalene. The bicyclic system of two fused seven-membered rings is known as heptalene (CCXIX); according to Craig's rule this hydrocarbon should be non-aromatic. Heptalene has been obtained as a viscous liquid, and is relatively unstable. It is stable in

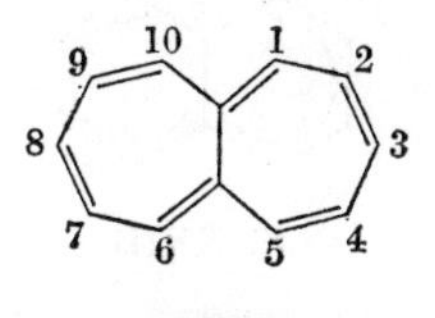

(CCXIX)

solution at 25° in the absence of oxygen, but polymerized in the presence of oxygen, or on warming. In dilute solution in carbon tetrachloride heptalene reacted instantaneously with bromine, and it was rapidly hydrogenated over platinum oxide. Its n.m.r. spectrum showed proton absorption only in the vinylic region and none in the lower-field aromatic region. Its ultraviolet absorption spectrum in cyclohexane showed maxima at 256 and 352 nm and a long tail throughout the visible region.

In the 1-heptalenium cation (CCXX), one ring becomes aromatic, being a tropylium ion. The n.m.r. spectrum is similar to that of the

1-azulenium ion, the proton absorptions of the aromatic ring being close to those of the tropylium ion. This heptalenium cation is non-planar, the *cis*-butadiene part of the molecule being twisted out of the plane of the tropylium moiety.

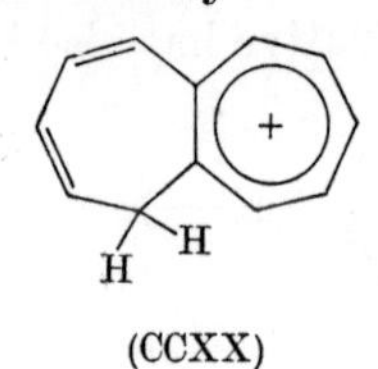

(CCXX)

4.4. Tricyclic and other systems.

Perinaphthenium cation. Perinaphthene (CCXXI) contains a saturated carbon atom, but the perinaphthenium cation (CCXXII) has increased possibilities of resonance. The high acidity of perinaphthene was noted in 1950, and the high calculated de-localization energy ($5{\cdot}83\beta$) for the cation has often been noted. The perinaphthenium cation has been successfully isolated and found to be thermodynamically stable; although very reactive it was satisfactorily stored under nitrogen. It reacted with water to give perinaphthene and perinaphthenone.

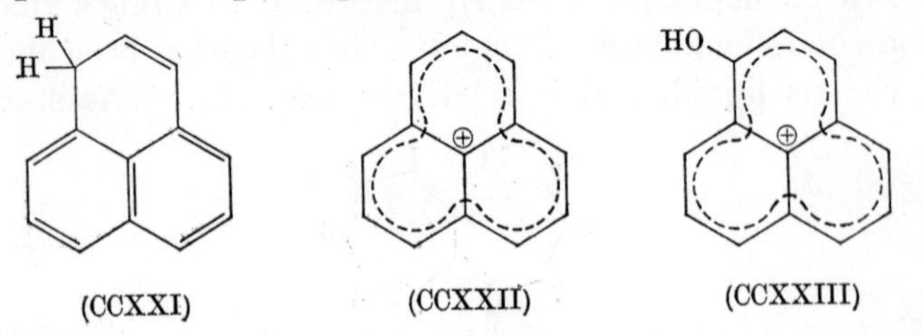

(CCXXI) (CCXXII) (CCXXIII)

Perinaphthenone is soluble in strong aqueous acids, presumably forming the hydroxy derivative (CCXXIII).

The perinaphthenide anion (CCXXIV) and perinaphthenyl radical (CCXXV) have also been prepared. The anion is readily oxidized by air to the radical. The radical was unstable in boiling benzene, but stable in an inert atmosphere in propylbenzene at 150°. The radical has been detected among the products of the pyrolysis of hydrocarbons at temperatures from 450 to 750°.

Pyrene isomers. Many non-benzenoid isomers of pyrene can be conceived; for example, the compound (CCXXVI) has been described. This is a stable hydrocarbon, being recoverable from solution in 70% perchloric acid. Its n.m.r. spectrum shows a multiplet for the nine ring protons at $\tau 1{\cdot}5$–$3{\cdot}3$, in addition to a singlet for the methyl protons at $\tau 7{\cdot}25$. The hydrocarbon forms a complex with *s*-trinitrobenzene.

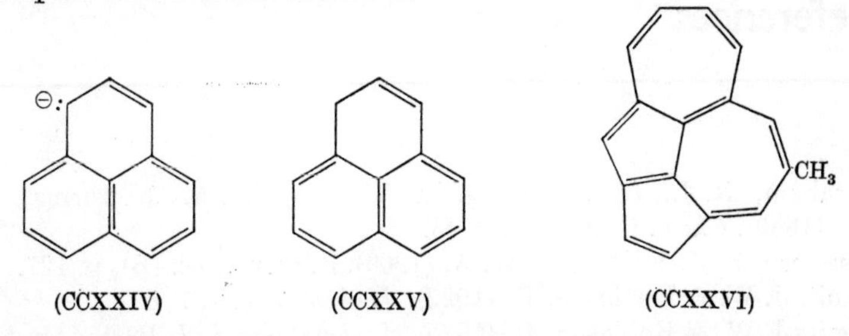

(CCXXIV) (CCXXV) (CCXXVI)

peri-*Cycloheptanaphthylene derivatives.* Preparations of *peri*-cycloheptanaphthylene (CCXXVII), *peri*-cycloheptacenaphthene (CCXXVIII) and *peri*-cycloheptacenaphthylene (CCXXIX) have been reported. All were found to exhibit aromatic character, particularly (CCXXIX), which failed to react with maleic anhydride at 80°. The blue acepleiadylene (CCXXX) behaves as a diene, and undergoes dimerization.

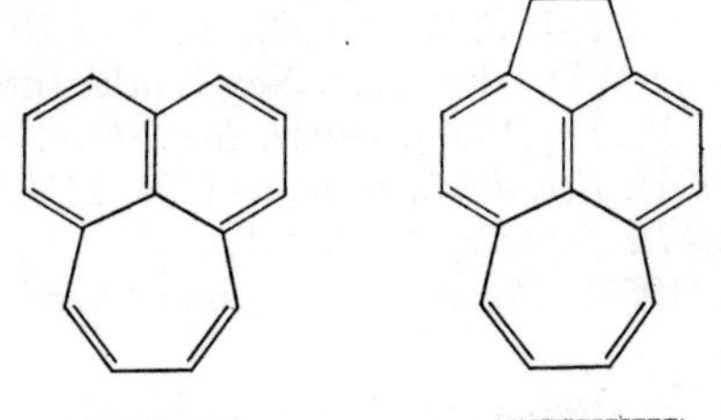

(CCXXVII) (CCXXVIII)

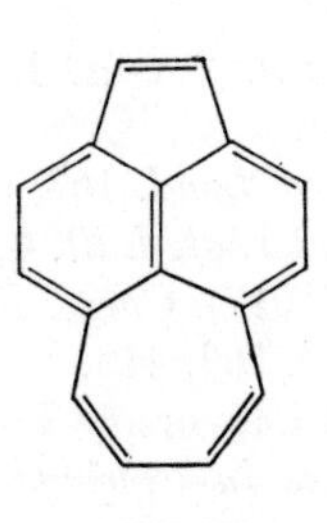

(CCXXIX)

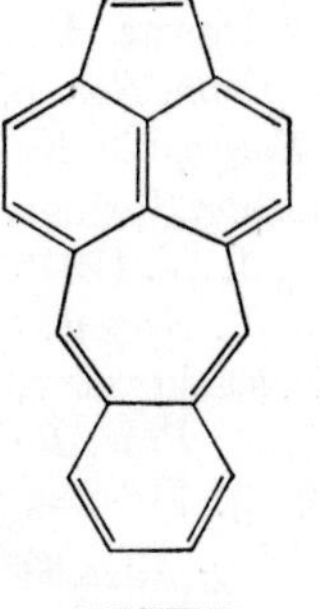

(CCXXX)

References

Abraham, R. J., Sheppard, R. C., Thomas, W. A. & Turner, S. (1965). *Chem. Comm.* no. 3, 43.

Abraham, R. J. & Thomas, W. A. (1966). *J. chem. Soc.* (B), p. 127.

Armit, J. W. & Robinson, R. (1922). *J. chem. Soc.* **121**, 827.

Armit, J. W. & Robinson, R. (1925). *J. chem. Soc.* **127**, 1604.

Badger, G. M. (1954). *The Structures and Reactions of the Aromatic Compounds*. Cambridge University Press.

Badger, G. M., Elix, J. A. & Lewis, G. E. (1966). *Austral. J. Chem.* **19**, 1221.

Badger, G. M., Lewis, G. E. & Singh, U. P. (1966). *Austral. J. Chem.* **19**, 257.

Bak, B., Christensen, D., Hansen, L. & Rastrup-Andersen, J. (1956). *J. chem. Phys.* **24**, 720.

Baker, W. & McOmie, J. F. W. (1959). In *Non-Benzenoid Aromatic Compounds* (edited D. Ginsburg). New York: Interscience.

Baker, W. & Ollis, W. D. (1957). *Quart. Rev. chem. Soc. Lond.* **11**, 15.

Bamberger, E. (1891). *Ber. dtsch. chem. Ges.* **24**, 1758.

Bastiansen, O. (1957). *Acta Cryst.* **10**, 861.

Bergmann, E. D. (1955). *Progress in Organic Chemistry* (edited J. W. Cook), **3**, 81.

Braude, E. A. (1945). *Ann. Rep. chem. Soc.* **42**, 123.

Bregman, J., Hirshfeld, F. L., Rabinovich, D. & Schmidt, G. M. J. (1965). *Acta Cryst.* **19**, 227.

Brown, R. D. (1950). *Trans. Faraday Soc.* **46**, 146.

Chung, A. L. H. & Dewar, M. J. S. (1965). *J. chem. Phys.* **42**, 756.

Clar, E. (1950). *J. chem. Soc.* p. 1823.

Clar, E. (1964). *Polycyclic Hydrocarbons*, 2 vols. London: Academic Press, and Berlin: Springer-Verlag.

Clar, E. & Wright, J. W. (1949). *Nature, Lond.* **163**, 921.

Collman, J. P. (1965). *Angew. Chem. internat. Edit.* **4**, 132.

Conant, J. B. & Kistiakowsky, G. B. (1937). *Chem. Rev.* **20**, 181.

Coulson, C. A. (1939). *Proc. Roy. Soc.* A **169**, 413.

Coulson, C. A. (1952). *Valence*. Oxford University Press.

Coulson, C. A. (1958). In *Steric Effects in Conjugated Systems* (edit. G. W. Gray). London: Butterworths.

Cox, E. G., Cruikshank, D. W. J. & Smith, J. A. S. (1958). *Proc. Roy. Soc.* A **247**, 1.
Cox, E. G., Gillot, R. J. J. H. & Jeffrey, G. A. (1949). *Acta Cryst.* **2**, 356.
Craig, D. P. (1959*a*). In *Non-Benzenoid Aromatic Compounds* (edit. D. Ginsburg). New York: Interscience.
Craig, D. P. (1959*b*). In *The Kekulé Symposium: Theoretical Organic Chemistry*. London: Butterworths.
Criegee, R. & Schröder, G. (1959). *Liebigs Ann.* **623**, 1.
Daudel, P. & Daudel, R. (1948). *J chem. Phys.* **16**, 639.
de Heer, J. (1954). *J. Am. chem. Soc.* **76**, 4802.
Dewar, M. J. S. (1945). *Nature, Lond.* **155**, 50.
Dewar, M. J. S. & Gleicher, G. J. (1965*a*). *J. Am. chem. Soc.* **87**, 685.
Dewar, M. J. S. & Gleicher, G. J. (1965*b*). *J. Am. chem. Soc.* **87**, 3255.
Dobler, M. & Dunitz, J. D. (1965). *Helv. Chim Acta* **48**, 1429.
Doering, W. von E. & DePuy, C. H. (1953). *J. Am. chem. Soc.* **75**, 5955.
Doering, W. von E. & Knox, L. H. (1954). *J. Am. chem. Soc.* **76**, 3203.
Dunitz, J. D., Orgel, L. E. & Rich, A. (1956). *Acta Cryst.* **9**, 373.
Elvidge, J. A. (1965). *Chem. Comm.* no. 8, 160.
Elvidge, J. A. & Jackman, L. M. (1961). *J. chem. Soc.* p. 859.
Emerson, G. F., Watts, L. & Pettit, R. (1965). *J. Am. chem. Soc.* **87**, 131.
Ferguson, G. & Robertson, J. M. (1963). *Adv. Phys. Org. Chem.* **1**, 203.
Fischer, E. O. (1955). *Angew. Chem.* **67**, 475.
Fitzpatrick, J. D., Watts L., Emerson, G. F. & Pettit, R. (1965). *J. Am. chem. Soc.* **87**, 3254.
Frisch, M. A., Barker, C., Margrave, J. L. & Newman, M. S. (1963). *J. Am. chem. Soc.* **85**, 2356.
Gerdil, R. & Lucken, E. A. C. (1965). *J. Am. chem. Soc.* **87**, 213.
Gerdil, R. & Lucken, E. A. C. (1966). *J. Am. chem. Soc.* **88**, 733.
Hanson, A. W. (1965). *Acta Cryst.* **18**, 599.
Heaney, H. (1962). *Chem. Rev.* **62**, 81.
Hückel, W. (1937). *Z. Elektrochem.* **43**, 827.
Ingold, C. K. (1938). *Proc. Roy. Soc.* A **169**, 149.
Ingold, C. K. (1953). *Structure and Mechanism in Organic Chemistry.* Ithaca, New York: Cornell University Press.
Jackman, L. M. (1959). *Applications of Nuclear Magnetic Resonance Spectroscopy in Organic Chemistry*. London: Pergamon Press.
Jaffé, H. H. & Orchin, M. (1962). *Theory and Applications of Ultraviolet Spectroscopy*. New York: Wiley.
Karle, I. L. (1952). *J. chem. Phys.* **20**, 65.
Kaufman, H. S., Fankuchen, I. & Mark, H. (1947). *J. chem. Phys.* **15**, 414.
Kaufman, H. S., Fankuchen, I. & Mark, H. (1948). *Nature, Lond.* **161**, 165.
Kealy, T. J. & Pauson, P. L. (1951). *Nature, Lond.* **168**, 1039.

Kekulé, A. (1865). *Bull. Soc. chim. Paris*, (ii) **3**, 98; *Bull. Acad. R. Belg.* **19**, 551.
Kekulé, A. (1866). *Liebigs Ann.* **137**, 129.
Kekulé, A. (1869). *Ber. dtsch. chem. Ges.* **2**, 362.
Klevens, H. B. & Platt, J. R. (1949). *J. chem. Phys.* **17**, 470.
Kofod, H., Sutton, L. E. & Jackson, J. (1952). *J. chem. Soc.* p. 1467.
Krebs, A. W. (1965). *Angew. Chem. internat. Edit.* **4**, 10.
Langseth, A. & Stoicheff, B. P. (1956). *Canad. J. Phys.* **34**, 350.
Longuet-Higgins, H. C. (1949). *Trans. Faraday Soc.* **45**, 173.
Longuet-Higgins, H. C. & Orgel, L. E. (1956). *J. chem. Soc.* p. 1969.
Longuet-Higgins, H. C. & Salem, L. (1959). *Proc. Roy. Soc.* A **251**, 172.
Longuet-Higgins, H. C. & Salem, L. (1960). *Proc. Roy. Soc.* A **257**, 445.
Miller, S. A., Tebboth, J. A. & Tremaine, J. F. (1952). *J. chem. Soc.* p. 632.
Nozoe, T. (1959). In *Non-Benzenoid Aromatic Hydrocarbons* (edited D. Ginsburg). New York: Interscience.
Ogryzlo, E. A. & Porter, G. B. (1963). *J. chem. Educ.* **40**, 256.
Pauling, L. (1931). *J. Am. chem. Soc.* **53**, 1367.
Pauling, L. (1959). In *The Kekulé Symposium: Theoretical Organic Chemistry*. London: Butterworths.
Pauling, L. (1960). *The Nature of the Chemical Bond*, 3rd ed. Ithaca, New York: Cornell University Press.
Pauling, L. & Brockway, L. O. (1934). *J. chem. Phys.* **2**, 867.
Pauson, P. L. (1955*a*). *Chem. Rev.* **55**, 9.
Pauson, P. L. (1955*b*). *Quart. Rev. chem. Soc. Lond.* **9**, 391.
Pauson, P. L. (1959). In *Non-Benzenoid Aromatic Compounds* (edited D. Ginsburg). New York: Interscience.
Pettit, R. (1960). *Tetrahedron Letters*, no. 23, 11.
Pink, R. C. & Ubbelohde, A. R. (1948). *Trans. Faraday Soc.* **44**, 708.
Prosen, E. J., Johnson, W. H. & Rossini, F. D. (1950). *J. Am. chem. Soc.* **72**, 626.
Pullman, B. & Pullman, A. (1952). *Les Théories Électroniques de la Chimie Organique*. Paris: Masson.
Raphael, R. A. (1959). In *Non-Benzenoid Aromatic Hydrocarbons* (edited D. Ginsburg). New York: Interscience.
Roberts, J. D., Streitwieser, A. & Regan, C. M. (1952). *J. Am. chem. Soc.* **74**, 4579.
Robertson, J. M. (1948). *Acta Cryst.* **1**, 101.
Sayre, D. & Friedlander, P. H. (1960). *Nature, Lond.* **187**, 139.
Selwood, P. W. (1956). *Magnetochemistry*, 2nd ed. New York: Interscience.
Sklar, A. L. (1937). *J. chem. Phys.* **5**, 669.
Sondheimer, F. (1963). *Pure and Appl. Chem.* **7**, 363.
Springall, H. D., White, T. R. & Cass, R. C. (1954). *Trans. Faraday Soc.* **50**, 815.
Stoicheff, B. P. (1954). *Canad. J. Phys.* **32**, 339.

St Pfau, A. & Plattner, P. A. (1936). *Helv. Chim. Acta* **19**, 858.

Streitwieser, A. (1961). *Molecular Orbital Theory for Organic Chemists.* New York: Wiley.

Sundaralingam, M. & Jensen, L. H. (1963). *J. Am. chem. Soc.* **85**, 3302.

Sundaralingam, M. & Jensen, L. H. (1966). *J. Am. chem. Soc.* **88**, 198.

Thiele, J. (1899). *Liebigs Ann.* **306**, 87.

Thiele, J. (1900). *Ber. dtsch. chem. Ges.* **33**, 666.

Thiele, J. (1901). *Ber. dtsch. chem. Ges.* **34**, 68.

Trotter, J. (1964). *Royal Institute of Chemistry, Lecture Series*, no. 2.

van Tamelen, E. E. & Pappas, S. P. (1963). *J. Am. chem. Soc.* **85**, 3297.

Waugh, J. S. & Fessenden, R. W. (1957). *J. Am. chem. Soc.* **79**, 846.

Wheland, G. W. (1955). *Resonance in Organic Chemistry.* New York: Wiley.

Wiberg, K. & Nist, B. J. (1961). *J. Am. chem. Soc.* **83**, 1226.

Wilcox, W. S. & Goldstein, J. H. (1952). *J. chem. Phys.* **20**, 1656.

Index